Salvador Yamil Limachi Limachi

Stochastic Modeling of Sedimentary Facies on a 3D Geological Grid

Salvador Yamil Limachi Limachi

Stochastic Modeling of Sedimentary Facies on a 3D Geological Grid

The art of estimating rock properties as a game of probabilities

ScienciaScripts

Imprint

Any brand names and product names mentioned in this book are subject to trademark, brand or patent protection and are trademarks or registered trademarks of their respective holders. The use of brand names, product names, common names, trade names, product descriptions etc. even without a particular marking in this work is in no way to be construed to mean that such names may be regarded as unrestricted in respect of trademark and brand protection legislation and could thus be used by anyone.

Cover image: www.ingimage.com

This book is a translation from the original published under ISBN 978-620-0-02462-6.

Publisher:
Sciencia Scripts
is a trademark of
Dodo Books Indian Ocean Ltd. and OmniScriptum S.R.L publishing group

120 High Road, East Finchley, London, N2 9ED, United Kingdom
Str. Armeneasca 28/1, office 1, Chisinau MD-2012, Republic of Moldova, Europe
Printed at: see last page
ISBN: 978-620-6-99473-2

GENERAL INDEX

ACKNOWLEDGMENTS

The present work could not have been carried out without the disinterested and valuable help of PhD.Udo Zimmermmann, master coordinator representing the cooperation of the University of Stavanger at UMSA.

Marco A. Montesinos M. career director and professors of the Petroleum Engineering career for the suggestions and support provided during the master's degree.

DEDICATION

This work is dedicated to my dear mother Brigida Limachi F., my aunts Luisa Limachi F. and Maxima Limachi A. for their support and encouragement to successfully complete the master's program.

EXECUTIVE OVERVIEW

The process of representing physical properties in a reservoir model consists of understanding its distribution (facies, porosity, permeability, etc.) in an adequate way, this requires the application of a series of criteria and techniques that aim to make a reliable and understandable model based on geological concepts.

The present work aims to analyze and apply all the steps required for successful modeling, including data integration, facies representativeness, and application of the appropriate technique for stochastic modeling; this sequence is crucial for an optimal representation. The points taken into account in the present study are:

- Identify sedimentary characteristics using seismic attributes (Coherence, Root Mean Square).

- Analyze electrical well logs (gamma ray, sonic, density, neutron, SP, resistivity, etc) in order to identify the vertical resolution of the model and understand the variation of properties around the wellbore.

- Correlating formation markers that indicate a geologic time in which a lithologic boundary is represented.

- Apply the stochastic modeling method, known as Sequential Indicator Simulation (SIS), which takes into account all the above points to define the parameters of its algorithm and thus calculate the corresponding facies based on the spatial probability with respect to the distance.

The sedimentary facies that are likely to be represented in this study are those that exceed the minimum resolution of the seismic record, i.e., greater than 10m according to the variation of the seismic wavelength, to this must be added the attenuation effect, which is a limitation of the reflection seismic, taking into account these limitations, we try to make a facies model for academic purposes.

CHAPTER I

INTRODUCTION

1.1. Introduction

Three-dimensional models play a very important role in the hydrocarbon industry, they are used to plan new wells, calculate reserves, and when linked to a flow simulator they help to predict production profiles. A key factor for professionals is the quantitative integration of 3D seismic, to obtain a better representation of reservoir properties between wells.[1]

The generation of 3D facies models is a crucial step, because it incorporates within itself much of the geological knowledge in a reservoir qualitatively and quantitatively. In this sense, it must respect geological laws (Walter's law) and observation laws, for facies successions that are consistent with each other and that present absence of significant erosion surfaces.[2]

Thus, models are becoming more and more complex, and are used to represent multiple properties such as facies, porosity, permeability, acoustic impedance, mechanical properties, dynamic properties (saturation and formation pressure) among others.

Geostatistics is a discipline that assists in the construction of models. Initially developed for applied purposes in the mining industry, geostatistics is a relatively new discipline, much of its development having taken place over the last forty years. The International Association of Mathematical Geology (IAMG) is responsible for developing the theoretical basis of geostatistics in 1968. In the following years from 1970 to 1980 very few people used geostatistical methods, however since that year the knowledge and the community of professionals using this technique has increased significantly.[3]

Thus, in 1994, the first volume of Stochastic Modeling and Geostatistics by authors Yarus and Chambers was published, which founded the practical application of the technique, improving the theory and contextualizing its application in a vital part of the reservoir

---

1 Ampilov, Y. (2010). Seismic Interpretation to Modelling and Assessment of Oil and Gas Fields. HOUTEN The Netherlands: EAGE Publications.
2 Le Blevec, T., Dubrule, O., Cedric, J., & Hampson, G. (2016). Building more realistic 3D facies indicator models. Bangkok: Paper presented at the International Petroleum Technology Conference.
3 Myers, D. E. (2006). Reflections on geostatistics and stochastic modeling. In T. Coburn, J. Yarus, & L. Chambers, Stochastic modeling and geostatistics: Principles, methods, and case studies (Vol. Computer Applications in Geology 5, pp. 11-22). AAPG American Association of Petroleum Geologists.

modeling process.[4]

Techniques such as krigging and conditional simulation are widely accepted for use in predicting reservoir properties between wells to create more realistic heterogeneous models.

The spatial distribution of facies with certain petrophysical characteristics, such as permeability, controls the way in which hydrocarbons are mobilized in the form of barriers or flow paths, and vertical heterogeneity in facies succession has been shown to impact the recovery factor of the field.[5]

It has recently become evident that stochastic (geostatistical) modeling not only provides distributions of geological parameters but also has a tremendous potential to integrate data coming from different sources. The branches that interact to provide useful input data are: geological knowledge (lithological and depositional modeling), geophysics, structural geology, petrophysics, and reservoir engineering. The possibility of integrating data at this level and at different scales makes stochastic modeling the most powerful technique currently available for reservoir characterization.[6]

The purpose of modeling sedimentary facies originating from a clastic environment in a geological 3D grid is to achieve the necessary representativeness on a model built on the producing formation or identified as a reservoir, this procedure uses as a basis geostatistical modeling in the form of weighting of the spatial distribution of facies (fine, medium and coarse grained). The initial way to identify geomorphologies and qualitative and quantitative facies indicators is using seismic attributes sensitive to sedimentological variations.[7]

1.2. Hypothesis

The use of seismic attributes (coherence, amplitude contrast) is the best way to identify sedimentary features associated with facies distribution, and their application improves the geological representation in the geostatistical modeling process with the SIS (sequential indicator simulation) method in a 3D geological grid.

4 Coburn, T., Yarus, J., & Chambers, L. (2006). Geostatistics and stochastic modeling: Bridging into the 21st century. In T. Coburn, J. Yarus, & L. Chambers, Stochastic modeling and geostatistics: Principles, methods, and case studies (Vol. Computer Applications in Geology 5, pp. 3- 9). AAPG American Association of Petroleum Geologists.
5 Zhou, Y., Muggeridge, A., Berg, C., & King, P. (2015). Quantifying cross flow and its impact on tertiary polymer flooding in heterogeneous reservoirs. European Symposium on Improved oil Recovery IOR.
6 Pelgrain, A., Cosetino, L., Cabrera, J., Jimenez, T., & Bellorin, O. (2002). Geologically oriented geostatistics: an integrated tool for reservoir studies. Mexico: Paper presented at SPE international Petroleum Conference.
7 Zakrevsky, K. E. (2011). Geological 3d Modelling. HOUTEN The Netherlands: EAGE Publications.

1.3. Target

Model sedimentary facies with the SIS (sequential indicator simulation) method on a geological 3D grid taking into account seismic attributes to obtain an adequate geological representation.

1.4. Specific objectives

*   Identify sedimentary features that serve as input data for modeling and correct representation in the 3D grid.

*   Proper selection of SIS (sequential indicator simulation) and variogram parameters.

*   Performing quality control of revenue data to maintain its distribution

*   Analyze the results obtained

1.5. Justification

*   A poor geological model leads to problems in the simulation, since it does not adequately represent the fluid flow in the reservoir, as a consequence, the production of a field can be overestimated or underestimated without being aware of it.

*   The limitation of resolution in reflection seismic means that stratigraphic hierarchies smaller than 10 meters cannot be taken into account; however, their presence must be taken into account by weighing their importance in the electrical records, this factor makes it important to understand the scale of the study.

*   Not all geological aspects are relevant to be represented in a geological model, that is why the representation objectives must be defined at the beginning of stochastic modeling. Correlation between the geological model and the representable one.

*   The best way to plan a well is to identify the most attractive locations in terms of hydrocarbon accumulation, the geological model is a qualitative and quantitative reference of attractive prospects. Identification of potential well sites

*   Representing well information in a geological model is very important since it references measurements close in distance to the well with a much larger scale (reservoir volume), the most important features (lithology, porosity, permeability, etc.) must be represented in the geological model such as shale laminations that constitute flow barriers in

the reservoir Integrating well information with the grid.

• An adequate geological model provides certainty regarding the field development plan, enabling adequate production management, as well as reserves monitoring.

1.6. Scopes

The present work has limitations typical of an academic work, where there are restrictions in access to information and in the perspective of analysis, being the following points the defined scopes:

No final computational comparisons are made between stochastic and deterministic modeling algorithms to quantitatively determine which method is best for the depositional system, instead the method is chosen based on qualitative criteria and general knowledge.

A data treatment is made based on the existing information, being that there could be referential values that are different such as: coordinate system, seismic reference data, or seismic that is not zero in displacement phase.

It does not question processes prior to stochastic modeling, i.e. it is assumed that the seismic interpretation, time-to-depth conversion and all required procedures are reliable within scientifically acceptable ranges.

The seismic tributes chosen are those for which we have specific knowledge of the algorithm performance, it is assumed that there may be other better techniques but the lack of practical basis restricts the use of them.

Facies interpretation is based on electrical logs that are sensitive to lithological representation in which facies classification is applied, and its variation is subject to input data.

1.7. Methodology

The working procedures for modeling properties are divided into three fundamental stages: [8]

• Construction of the geometric volume from horizons and faults, which were interpreted from seismic data.

• Construction of the stratigraphic model that provides a set of cells to be used in the

---

8 Dubrule, O. (2003). Geostatistics For Seismic Data Integration In Earth Models. HOUTEN The Netherlands: EAGE Publications.

assignment of petrophysical properties.

- Modeling of properties using well information, i.e. that derived from electrical logs.

In this sense, the data required for the construction of property models are as follows:

- Horizon and fault data. Which are regular surfaces that define top and base of reservoir intervals. Horizons are obtained from 3D seismic interpretation and then converted to depth. Faults are those interpreted continuous surfaces. These data are fundamental for the construction of geological volumes.

- Electrical logs and formation markers. The fundamental logs can be those indicators of lithology or petrophysics, these measurements provide information throughout the well trajectory. Formation markers are spatial points where there is lithological information about the formations crossed by the well.

- Volumes of seismic attributes. Which are qualitative indicators of sedimentary bodies, and from which information can be extracted.

The methodology used in this work takes into account the following steps:[9]

- Review of all data obtained. By means of physical and spatial verification in the files, that is to say that they are susceptible to import and use in the work.

- Bibliographic review. In order to collect geological and field information.

- Identification of representative data. Selection of those initial data (seismic or electrical records) or calculated data (facies) that are useful as input data in the calculation process.

- Attenuation of electric logs. So that the values of the logs are assigned to the stratigraphic cells of the model in the well trajectory, these must be attenuated, due to the variation of resolution, the type of attenuation depends on the property and goes from arithmetic to more advanced weightings, methods of representativeness control must be used with the use of histograms and verification of values, to avoid exaggerations that eliminate valuable information.

- Spatial continuity analysis. It refers to the analysis of variograms to characterize the

---

9 Zakrevsky, K. E. (2011). Geological 3d Modelling. HOUTEN The Netherlands: EAGE Publications.

spatial continuity in the variation of reservoir properties, these calculations are performed as a function of stratigraphic distances, the analysis can be done vertically and horizontally. Since the estimation of the vertical resolution of the geological model is a function of the variogram based on the electrical logs of lithology (gamma ray, density, sonic), we can understand the maximum range of variation of properties at distances close to the well. Identify the horizontal orientation and trend by constructing horizontal variograms, the maximum variation allowed should not exceed 1/2 of the range in the smaller horizontal variogram, in the seismic attribute that has the highest sedimentological representativeness.

•	Classification methods will be used to identify spatial variations of facies, according to the sedimentological process identified as predominant and/or most influential in the interval of interest.

•	Perform sensitivity calculations to assess the impact on the entire distribution, taking into account that the validation of the parameters of influence requires a high level of expertise

CHAPTER II

IDENTIFICATION OF SEDIMENTARY BODIES

Most of the procedures for the recognition of sedimentary bodies take 3D seismic data. In the first instance the interpreter performs a scan in-line, cross-line and horizontally, the objective in this first instance is to identify features that are anomalous which usually include amplitude bright spots, abrupt lineaments, or features that are geologic to the naked eye. Another way involves the application of opacities in order to visualize specific amplitude values that allow a paleomorphology to be seen. Once these characteristics are identified, we proceed to interpret amplitudes in the reflector of interest, extract amplitudes in a defined interval, analyze seismic attributes in an interval such as coherence, RMS, acoustic impedance, among others.

At the conclusion of these steps, the meaning of these patterns must be geologically defined, which can be for example fluvial or deep water channels, landslides, landslides, continental shelf boundaries, carbonates, etc.

A critical step in the evaluation of any seismic characteristic is that this expression must be scientifically reasonable in multiple directions, as well as in the different seismic planes of view. A complement is also referred to the available borehole data, which provide lithological and sedimentological knowledge. Analogy data with known current or ancient environments can be used to enhance the credibility of the interpretation.[10]

2.1. Identification by seismic attributes

A seismic attribute is any operation performed on seismic data that helps to enhance the visualization or quantify features that are of interest in the interpretation zone, seismic attributes can be of time, amplitude, phase and seismic frequency. A good seismic attribute must be directly sensitive to the desired geologic feature or reservoir property of interest and thus allow us to define structural or depositional features and infer properties of interest. Seismic attributes are widely used in seismic interpretation, Fig 2.1 illustrates in a practical way the importance of applying seismic attributes to highlight depositional features that are amenable to three-dimensional seismic visualization.

---

10 Posamentier, H., & Allen, G. (1999). Siliciclastic Sequence Stratigraphy-Concepts and Applications. Tulsa, Oklahoma, U.S.A: SEPM Society for Sedimentary Geology.

In Fig 2.2 we can see an example of a coherence attribute that highlights the limits of a channel, in which sinuous elements can be differentiated that in the case of element 1 is older and in element 2 is newer, for a) applying the attribute and highlighting characteristics, in b) identifying the seismic facies according to their seismic expression, in c) characterizing the geologic body. Because it highlights textural characteristics, it can be used in different types of stratigraphic and sedimentological analysis in the interval of interest, considering that they represent conceptual relationships between seismic facies and their deposition, the environment can be understood in a three-dimensional perspective outlining its depositional characteristics and geologic properties.[11]

Fig. 2.1 Example of use of seismic attribute

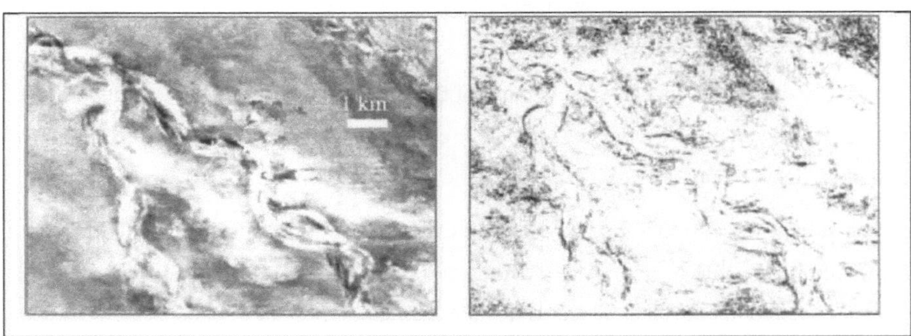

Source: (Chopra & Marfurt, 2007)

Fig. 2.2 Geological characterization by seismic attribute

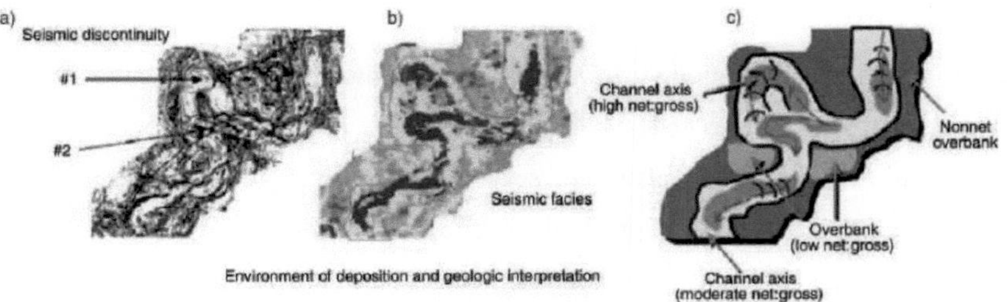

Source: (Catuneanu, 2006)

All the techniques that are applied for this purpose contribute to a better understanding of a particular geomorphologic feature. As in many aspects of seismic interpretation, it is

---

11 Posamentier, H., & Allen, G. (1999). Siliciclastic Sequence Stratigraphy-Concepts and Applications. Tulsa, Oklahoma, U.S.A: SEPM Society for Sedimentary Geology.

imperative to be aware of the errors and deviations that the model entails at the time of construction, so the interpreter must be able to distinguish between depositional expressions and deformations by geophysical processing. Another aspect to consider is the limitations of the vertical resolution, according to the geological context and the way in which the attributes used were calculated.[12]

While the attributes help to identify sedimentary bodies, knowledge of how the algorithm works is vital to determine the most important variables, highlight as many variations as possible and visually obtain something that can be recognized geologically.

10 1.1. RMS (Root mean square)

RMS whose meaning translated is the average of square roots of the seismic trace which is a function of time, this attribute is related to the energy contained in the trace, and its calculation is performed within a moving calculation window. The formula of RMS amplitude is:

$$RMS\ (t) = \sqrt{\frac{1}{N}\sum_{k=-\frac{N}{2}}^{\frac{N}{2}}{}_N[f(t+k)]^2} \quad\ldots\ldots\ldots\ldots Ec.2.1$$

The RMS operator is inherently sensitive to individual strong reflectors, because this feature is characteristic of sedimentary bodies that have a relatively high impedance contrast, the final calculation becomes sensitive to sedimentary parameters.

The RMS attribute is a post-stack attribute and can be used to directly identify hydrocarbons in an area of interest. However, it is sensitive to noise since it squares all values within a calculation window. RMS is a smoother version in terms of reflection stress estimation, it is applied in the same way to visualize amplitude anomalies in seismic, the resolution can be modified by changing the calculation window to produce a smoother amplitude estimate, it helps in the identification of facies containing coarse grains of high porosity, compaction effects such as marble and siltstone, and also unconformities; as seen in Fig. 2.3, the application of this attribute to seismic amplitude anomalies can be used for the identification of hydrocarbons in a seismic zone of interest. 2.3 the application of this attribute highlights

12 Doyen, P. (2007). Seismic Reservoir Characterization An Earth Modelling Perspective (Vol. Education Tour Series). Houten The Netherlands: EAGE Publications.

specific features in the seismicity.[13]

Fig. 2.3 Application of RMS in seismic.

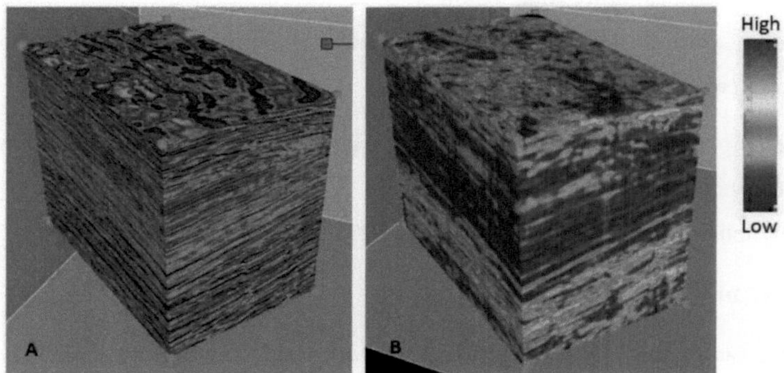

Source: (Al-Anezi, Kumar, & Eibad, 2013).

2.1.2.   Coherence

The coherence attribute is defined as a measure of the similarity between wavelets. When looking at a processed seismic section we observe characteristics that are the product of the convolution of the wavelet with the subsurface geology, this response changes in terms of amplitude, frequency and phase depending on the acoustic impedance contrast and the thickness of the strata, the concept of coherence can be better visualized in the continuity of reflectors in Fig.2.4 where it is shown from top to bottom coherent events defined in lateral continuity, each of these seismic responses corresponds to geological expressions in the section and their expression goes from continuous, discontinuous to chaotic reflectors.[14]

Fig.2.4 Expression of seismic waves

---

13 Chopra, S., & Marfurt, K. (2007). Seismic Attributes for Prospect Identification and Reservoir Characterization (Vol. Geophysical Developments Series No. 11). Tulsa, OK U.S.A: SEG Society of Exploration Geophysicists.
14 Chopra, S., & Marfurt, K. (2007). Seismic Attributes for Prospect Identification and Reservoir Characterization (Vol. Geophysical Developments Series No. 11). Tulsa, OK U.S.A: SEG Society of Exploration Geophysicists.

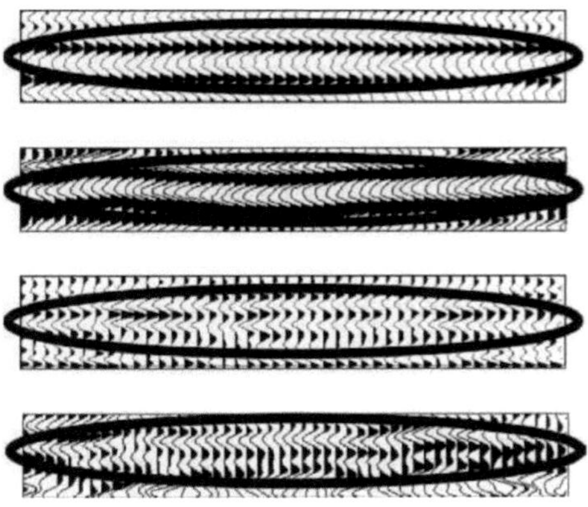

Source: (Chopra & Marfurt, 2007)

When you have 3D seismic volumes you can appreciate spatial evolutions of the structure and its stratigraphy, as long as the quality of the seismic acquisition is adequate for that, for example depositional systems (channels or meanders) are better visualized in horizontal slices. Usually the results of this technique provide valuable input data to perform modeling of properties in the studied reservoir, this implies that its use reduces exploration risks, helps in the better understanding of reservoir production.

The coherence algorithm uses a spatial search window for comparison with neighboring wavelets, as shown in Fig.2.5 where the central wavelet is the comparison root for the neighbors, in a) the possible comparison wavelets and in b) the crossline and inline compared wavelets.

Fig.2.5 Example of coherence algorithm search

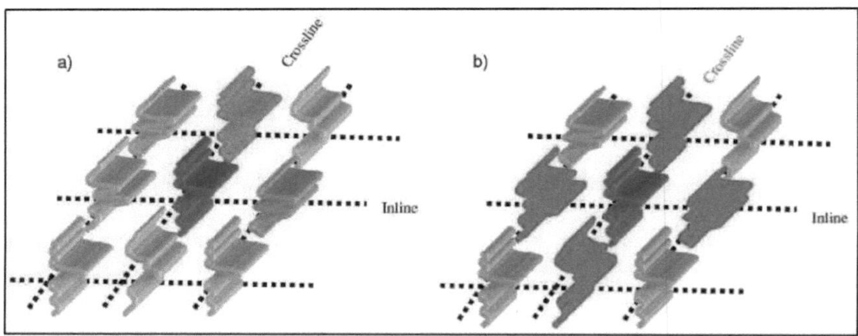

Source: (Chopra & Marfurt, 2007)

A variant of the algorithm is based on the calculation of variance across neighboring wavelets according to the equation:

$$var(t,p,q) = \frac{1}{J}\sum_{j=1}^{J}\left[u_j(t - px_j - qy_j) - \langle u(t,p,q)\rangle\right]^2 \quad \ldots\ldots\text{Ec.2.2}$$

Where the expression $\langle u(t,p,\text{q})\rangle$ is defined as

$$\langle u(t,p,q)\rangle = \frac{1}{J}\sum_{j=1}^{J}u_j(t - px_j - qy_j)\ldots\ldots\ldots\text{Ec.2.3}$$

Thus the above equations are considered as a quantification of how well the wavelet trace resembles or fits an average trace within the analysis volume[15] . The final application of the algorithm on a seismic record can be seen in Fig.2.6.

Fig.2.6 Application of the coherence algorithm

Source: (Posamentier & Allen, 1999)

2.2. Identification by means of electrical records

In a depositional analysis, such as a paleo-delta, gamma logs are likely to aid in sedimentological analysis and thus determine the nature of contacts and details of facies relationships. These studies allow to evaluate the degree of compartmentalization in the

15 Posamentier, H., & Allen, G. (1999). Siliciclastic Sequence Stratigraphy-Concepts and Applications. Tulsa, Oklahoma, U.S.A: SEPM Society for Sedimentary Geology.

various stages of the field life from exploration to production, the main benefit being to provide a large scale facies distribution, as can be seen in Fig.2.7 where the regression of a deltaic depositional system is illustrated, where GR=gamma ray; CH= fluvial channel; CS= crevasse splay; MFS= màximum flooding Surface. It should be noted that the maximum flooding surfaces are associated with finer grained sediments and their position indicates the overall geometry of the progradation, in this case the reservoir includes at least five hydrodynamic flow units, each corresponding to a stage of delta progradation.[16]

Electrical logs provide information of the physical properties of the rocks, but not of the lithology in a direct way, being indirect measurements in nature. The most commonly used logs in the interpretation of siliciclastic successions in lithological terms are: spontaneous potential and gamma ray, so the interpretation is made in terms of decreasing and increasing grain size. Consequently, high gamma ray values may correspond to a variety of sedimentation conditions, from shelf, deep water, coastal plain or lacustrine environments. However, these high gamma ray values are indicative of restricted periods of bottom current circulation and/or times of reduced sediment input.

It is also important to note that, when it comes to geological interpretation of electrical logs, different depositional systems produce similar shapes, therefore correct interpretation requires the integration of multiple studies and data including: core analysis, mud logging rock cuttings, biostratigraphy, and seismics. An example is seen in Fig.2.8 where comparisons of block-type sandstone responses are similar for fluvial, estuarine, beach, shallow marine, and deep marine environments, where 1=filled fluvial channel, 2=filled estuarine channel, 3=coastal deposits, 4=deep water channel filled with turbidites, 5=beach deposits.

Fig.2.7 Geological interpretation of electrical records

---

16 Catuneanu, O. (2006). Principles of Sequence Stratigraphy. Amsterdam, The Netherlands: Elsevier.

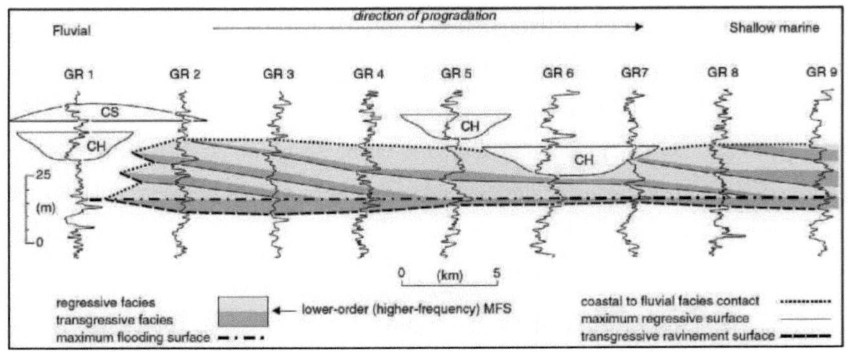

Source: (Catuneanu, 2006)

Fig.2.8 Example of block shape in gamma ray

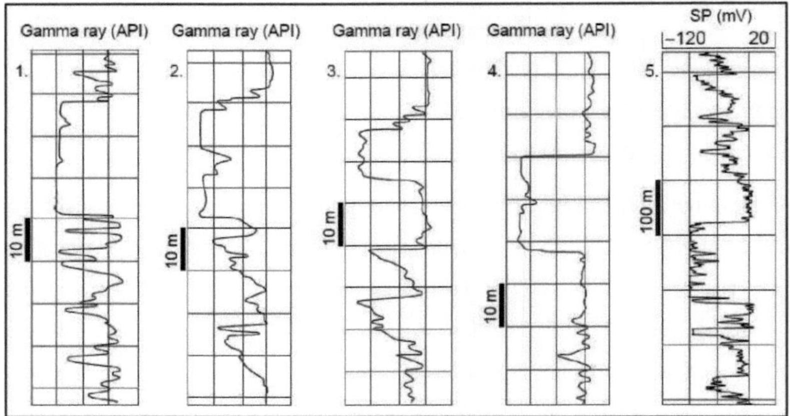

Source: (Veeken & Van Moerkerken, 2013).

Similarly, jagged forms in logs cannot be diagnosed for a particular depositional environment, however what they indicate is a fluctuation of energy conditions prevailing in the deposition of fine and coarser sediments (heterolithic facies), these conditions can be reached in non-marine, marginal marine and deep marine environments. An example is given in Fig.2.9 where the tracks in the examples are 1=river system, 2=deltaic plain, 3=continental shelf above storm level, 4=deep water continental shelf or basin base respectively.[17]

Fig.2.9 Example of jagged expression in gamma ray

---

17 Veeken, P., & Van Moerkerken, B. (2013). Seismic Stratigraphy and Depositional Models. HOUTEN The Netherlands: EAGE Publications.

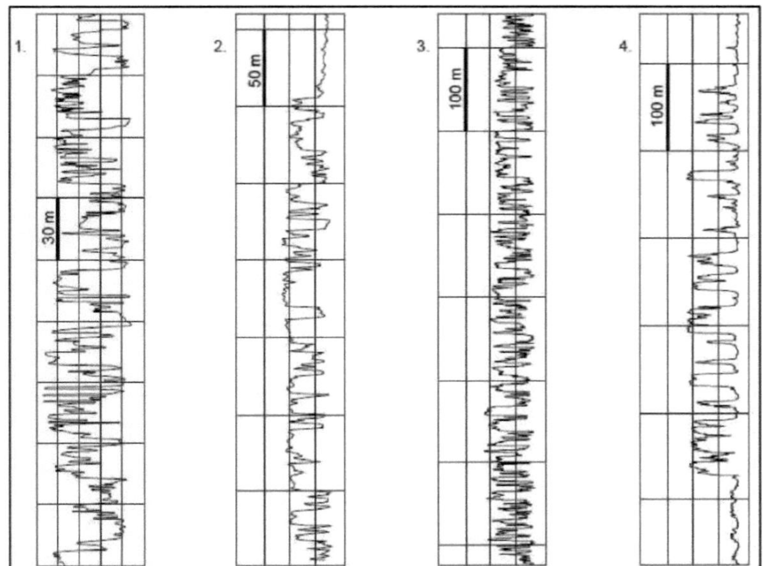

Source: (Veeken & Van Moerkerken, 2013).

Monotonic successions with little erratic behavior, dominated by fine-grained sediments may be common in many environments, including deep-water overlaps (areas of seafloor located outside the flood zone of a channel) as seen in Fig. 2.10 where fine-grained successions exist, where 1=deep seafloor basin, 2=shore below storm level but above the shelf. It should be noted that the clay bodies overlie shallower sand deposits. The association of records provides important clues for the interpretation of paleodepositional environments, the arrow indicates a coastal flooding event.[18]

Fig.2.10 Example of monotonic succession in registers

18 Veeken, P., & Van Moerkerken, B. (2013). Seismic Stratigraphy and Depositional Models. HOUTEN The Netherlands: EAGE Publications.

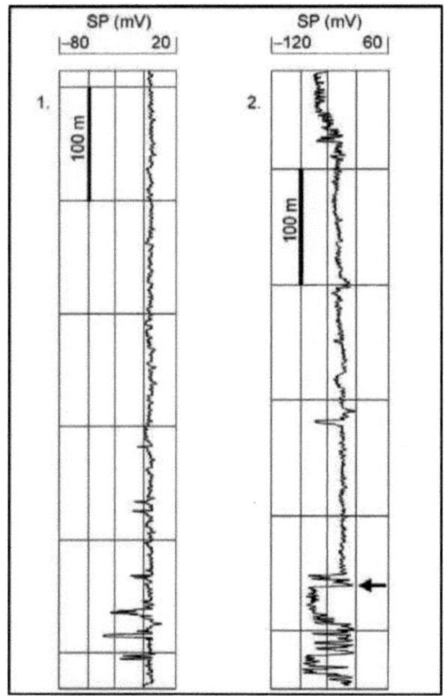

Source: (Veeken & Van Moerkerken, 2013).

Finally, the sequential patterns of increasing grain size indicate a gradual change towards a depositional environment that has a high energy level, but in a progressive manner. Fig.2.11 visualizes the expression in the environments in which this type of pattern can be found, which include for 5-30m: deltaic environments where sandy bodies progressing in the distributary, progradation of shallow shore deposits where there is openness. For less than 5m: flood bodies (crevasse splay) in fluvial environments, also in gravitational flow systems in deep environments especially at the distal limits of a turbidite lobe where CH= channel, CS= crevasse splay.[19]

Fig.2.11 Example of sequential pattern in registers

19 Veeken, P., & Van Moerkerken, B. (2013). Seismic Stratigraphy and Depositional Models. HOUTEN The Netherlands: EAGE Publications.

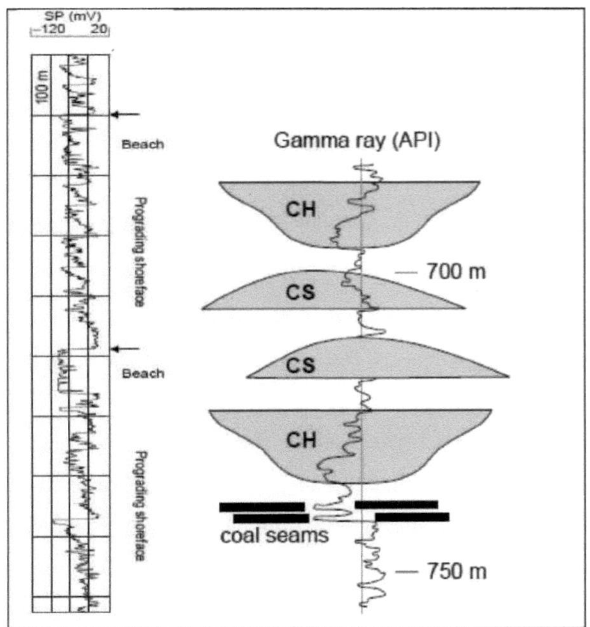

Source: (Catuneanu, 2006)

CHAPTER III

CRITERIA IN FACIES MODELING

Facies modeling belongs to the branch of property modeling and is responsible for giving values to each of the cells (cell grids) with discrete or continuous properties, where all available geological information must be used to build a model that is as close as possible to reality.

The fundamental requirement is to have a geological knowledge, notion of the reservoir connectivity and the level of heterogeneity of the facies, the input data must be such that the descriptive information of the facies is understandable, in terms of shape, size and orientation of the sedimentary bodies, in Fig.3.1 you can see the different sedimentary environments in which facies modeling can be applied in direct relation to hydrocarbon reservoirs. Sedimentary rocks that are formed from material coming from previous rocks (detrital) can have a depositional character that can be lacustrine, continental, shallow or deep water, the depositional environment determines the characteristics of the reservoir.

The depositional environment is a vital part of the evaluation of a well and a field, thanks to this we can define the prevailing lithology, and have criteria regarding the variation over short or long distances, for example a delta in progradation on the continental shelf tends to present conglomerates, while if it is deep water it is more common to find very fine-grained shales.[20]

The objective of facies modeling is to capture the reservoir architecture with flow units and barriers. Flow units can be modeled as zones and barriers can be modeled as faults. Depending on the depositional environment and the amount of input data the most suitable algorithm can be chosen.[21]

Fig.3.1 Depositional environments

---

20 Veeken, P. C. (2007). Seismic Stratigraphy, Basin Analysis and Reservoir Characterization (Vol. Seismic Exploration Volume 37). The Netherlands: Elsevier.

21 Pyrcz, J., Gringarten, E., Frykman, P., & Deutsch, C. V. (2006). Representative input parameters for geostatistical simulation. In T. Coburn, J. Yarus, & L. Chambers, Stochastic modeling and geostatistics: Principles, methods and case studies (Vol. Computer Applications in Geology 5 volume II, pp. 123-137). AAPG American Association of Petroleum Geologists.

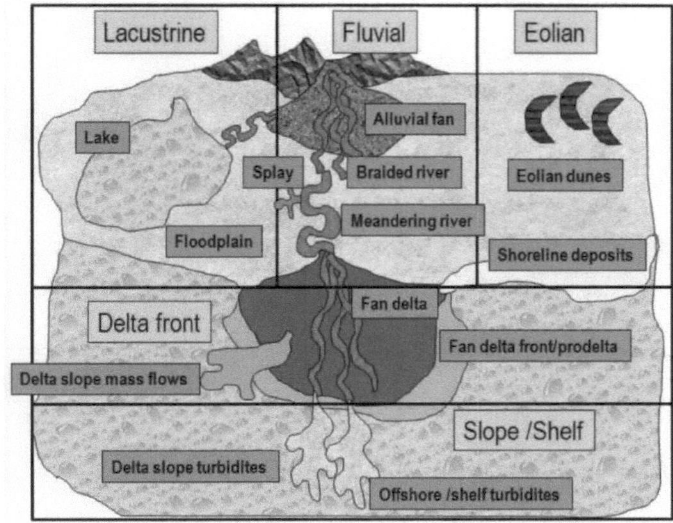

Source: (Le Blevec, Dubrule, Cedric, & Hampson, 2016).

3.1. Algorithm selection

For the process of facies modeling there are many techniques that are useful under different circumstances, for the selection of the algorithm it is important to know the basic concepts of application, in Fig.3.2 you can see the classification and the family to which these methods belong as well as the ideal examples of the results of each of these algorithms.[22]

Fig.3.2 Available facies modeling methods

Deterministic			Learning systems
Estimate	Direct assignment		Artificial
Indicator Kriging	Assign values	Interactive	Neural Net

---

22 Coburn, T., Yarus, J., & Chambers, L. (2006). Geostatistics and stochastic modeling: Bridging into the 21st century. In T. Coburn, J. Yarus, & L. Chambers, Stochastic modeling and geostatistics: Principles, methods, and case studies (Vol. Computer Applications in Geology 5, pp. 3- 9). AAPG American Association of Petroleum Geologists.

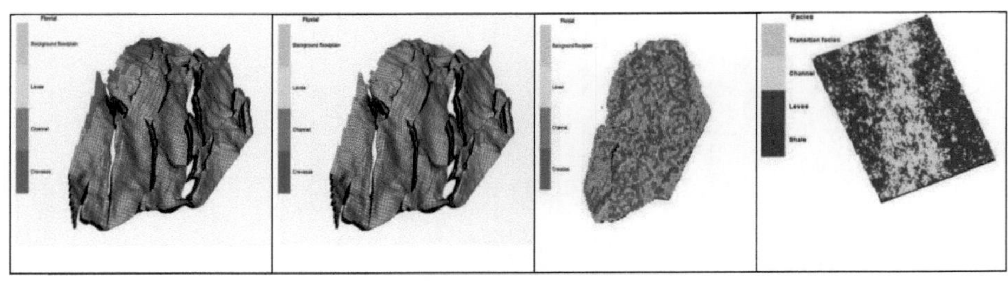

Stochastic				
Pixel based				Object based
Sequential Indicator Simulation	Tuncated Gaussian Simulation	Truncated Gaussian Simulation with Trends	Multipoint Facies Simulation	Object Modeling

Source: (Zakrevsky, 2011)

3.1.1. Deterministic and artificial algorithms

Deterministic algorithms are those that will always give the same result with the same input data, these algorithms are easy to understand since the assignment of properties to a cell is assigned a definite, logical and easy to predict value. Their disadvantage is that the results will be very stylized or smooth in form, although evidence and experience suggests otherwise. Based on this consideration it is not possible to have any notion of the uncertainty of the model in the distances of the input data. The use of these techniques occurs when there is a large amount of input data, i.e. many boreholes and high resolution 3D seismic. According to Fig.3.2 the description of deterministic methods and algorithms is as follows: (Doyen, 2007)

Estimate

- Indicator krigging allowing a property distribution by enforcing a predefined histogram

Direct assignment

- Assign values which allows you to assign values on the basis of functions

- Interactive allows painting facies directly on the 3D model

Learning systems

- Neural nets using a classification model from a learning process and variable assignment for the creation of a discrete property

3.1.2.   Stochastic algorithms

Stochastic methods are useful when it comes to incorporating secondary information, such as seismic data, for their calculation using a random source in addition to input data, so it is important to note that multiple calculations with the same parameters give similar but not equal results. These algorithms are much more complex and take much longer to perform, however they respect more aspects of the input variables in relation to their continuity and variability. This means that high and low values will appear in the results, and their distribution will resemble reality with Auctuations close to the close average. These algorithms generate aspects that have the character of a Random function distribution, so that the uncertainty analysis of these effects can be quantified in the realizations. These techniques are used in conditions where the input data are scattered and their prediction improves as the input data density increases. These methods allow to obtain results that are equally likely under the same input parameters. According to Fig.3.2 the description of the stochastic methods and algorithms is as follows:[23]

Object based

- Object modeling allows the assignment of a discrete facies model with different bodies of varying geometry, a facies code, and fraction

Pixel based

- Sequential indicator simulation (SIS) allows a stochastic distribution of the property

---

23 Dubrule, O. (2003). Geostatistics For Seismic Data Integration In Earth Models. HOUTEN The Netherlands: EAGE Publications.

using a predefined histogram, directional variables such as variograms and extensional trends are fulfilled.

• Truncated gaussian simulation TGS is a fast discrete property modeling technique, where facies are known to follow a sequential behavior and where the variograms for each facies are the same, usually used in carbonate environments, it can handle a large amount of input data, specified fractions, and trends in different planes.

• Truncated gaussian simulation with trends is a technique that uses depositional trends in maps where sequential relationships are defined to estimate the best approximation in the facies to be modeled, once this step is completed it becomes a TGS type algorithm.

• Multi point facies simulation (MPFS) is an algorithm based on multiple point statistics and creates models that resemble model patterns, matching their shape to the spatial correlation between multiple points at the same time, this method does not use variograms and replaces them with pattern images.

3.1.3. Selection

For the practical application in the present project, the need arises to find the most suitable method with which to achieve the best results. In this sense, the following points are considered for the adequate selection:

• Stochastic methods are generally more flexible than deterministic methods, for example with the use of Indicator krigging you will never get different results.

• Stochastic methods handle trends better and if you need to repeat the calculations exactly, you can do it with the same root value.

• In practice, it is impossible to perform uncertainty analysis using krigging.

• The choice also depends on the conceptual model of the reservoir and the experience of the expert modeler.

• The use of object modeling may be reasonable in the case of river beds that are reservoirs, but taking into account that the parameters of the geobodies may be chosen in such a way that there may be a false match or an error in the calculation.

• Direct assignment methods are recommended for small corrections but not for complete assignments and require a high level of experience and expertise to understand the consequences of each change made.

It is concluded that a flexible method that incorporates secondary variables from seismic data is needed for the project, so a method that incorporates secondary variables from seismic data is chosen.

stochastic. Since these turbidites cannot be defined directly but in an approximate way, a pixel based method will be used. Among the available algorithms, Sequential Indicator simulation was chosen because of the high variability in the geobodies and also because there are no facies distribution maps to which we have access.

3.2. Stochastic modeling

Anisotropy is a very important factor to consider in stochastic modeling since it is a way to measure how the data values change in a preferred direction Fig.3.3 where it can be seen that the variation of grain size distribution in the channel transversely is much larger than in the flow direction of the channel. When deciding on a modeling process one must consider: the distribution of facies values, variability of values, connectivity of extreme values, and our ability to assess the impact of uncertainty.[24]

Fig.3.3 Depositional system with directional variation

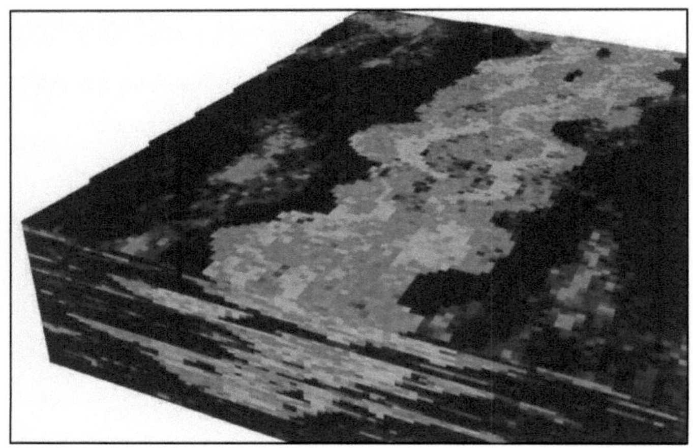

Source: (Pelgrain, Cosetino, Cabrera, Jimenez, & Bellorin, 2002).

---

24 Dubrule, O. (2003). Geostatistics For Seismic Data Integration In Earth Models. HOUTEN The Netherlands: EAGE Publications.

3.2.1. Variogram

Variograms are used to model how two values in space or time are correlated. In general, two values that have spatial proximity tend to be more similar than two values that are more distant in dimensional scale. Statistics based on a single variable cannot take this aspect into account, i.e. two distributions may have the same mean and variance, but the way they are correlated with each other cannot be defined.[25] A variogram can be displayed in a variance vs. class distance plot, the constituent parts of a variogram are Fig.3.4:

• Variance: a measure of how different the members of a data collection are from each other.

• Lag: separation distance between data points or measurements.

• Sill: variance at a point where the trend becomes horizontal and describes the variation between two uncorrelated samples.

• Range: correlation distance beyond which the points of the collection do not present any statistical similarity, in other words, it describes the separation distance beyond which there is no change in the degree of correlation between pairs of data in the collection and beyond which two points have only a random or any relationship.

• Nugget: degree of non-similarity at a distance corresponding to zero, which describes the variation of the data over a short distance, and is best identified vertically where the data interval is very small, it is also interpreted as small scale variation and has influence when thin laminations exist, which can cause rapid changes in porosity over short distances.

Fig.3.4 Typical variogram and its parts

25 Myers, D. E. (2006). Reflections on geostatistics and stochastic modeling. In T. Coburn, J. Yarus, & L. Chambers, Stochastic modeling and geostatistics: Principles, methods, and case studies (Vol. Computer Applications in Geology 5, pp. 11-22). AAPG American Association of Petroleum Geologists.

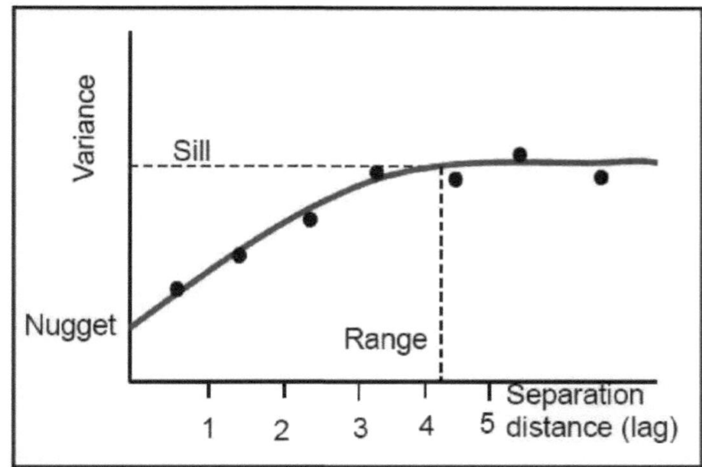

Source: (Coburn, Yarus, & Chambers, 2006)

The essential information provided by the variogram serves to identify the relationship in which two points that are close to each other tend to be more similar (small variance for small distances).

As with any data analysis, variogram models can be estimated for comparison with the data in the collection.

There are three types of variogram models:[26]

• Spherical Variogram: it is the simplest and has a linear behavior in short distances with abrupt transitions in the sill limit, they are the ones that respond better and are more stable in convergence when solving the system of krigging equations, the influence of the data is limited by the range.

• Exponential Variogram: it presents an erratic behavior at short distances with asymptotic approximations at distances larger than the sill.

• Gaussian Variogram presents a high level of continuity at short distances and then presents exponential behavior at longer distances, usually it tends not to be used in discrete data and should be used with caution for permeability.

A variogram map Fig.3.5 is a way of presenting individual variograms that were calculated

---

26 Dubrule, O. (2003). Geostatistics For Seismic Data Integration In Earth Models. HOUTEN The Netherlands: EAGE Publications.

in different directions, clearly indicating the direction in which anisotropy exists through an oval shape, where the center represents zero in search distance, and the contours the direction of greatest and least anisotropy with respect to the analyzed data.

Fig.3.5 Area Variogram

Source: (Le Blevec, Dubrule, Cedric, & Hampson, 2016).

The construction of variograms is very useful for:

•   Use geostatistical algorithms, being the primary input function for estimating values, e.g., if krigging is used, a smooth and continuous variogram that meets specific mathematical properties is required, a model that approximates the experimental variogram can meet these conditions.

•   A vertical variogram is important for determining optimal stratigraphic thicknesses, since it is a good indicator for defining strata thickness increments in a particular zone, since it is desired to represent strata where facies differences can be easily visualized.

•   Determine directions and degree of anisotropy by constructing horizontal variograms.

3.2.2.   Kriging

Kriging is a computational algorithm for the estimation of a linear system of equations where the values of the variogram are known (variance and average) and where the estimation parameters are unknown.

Krigging makes assumptions about the input data, and the behavioral form of the variables being analyzed, which means that the property must behave consistently with the chosen volume, i.e., it must not have external trends, and if it does, they must be corrected before modeling.[27]

The results obtained with the algorithm can be summarized in two principles:

• The first, referring to the fact that a datum that is close in geological distance to the unknown value must have a high weighting.

• The second, referring to data that are close to each other must share the same weighting.

In conclusion, practically speaking, the estimated value that minimizes the variance error is krigging and the result is a probability distribution that varies around the input data, it is an estimation technique based on the interaction between the variogram parameters and the neighboring data.

In general krigging is an average motion algorithm where the objective is to calculate the parameters as seen in Fig. 3.6. Being a distribution on a surface, the only known variable is at the well location and where the values of the other variables must be estimated, the base krigging equation is:

$$Z(x_0) = \Sigma_i^n \lambda_i \, Z(x_i) \, \ldots \ldots Ec.3.1$$

Where:

$Z(x_0)$ are unknown values that are weighted based on the sum of known values and calculating the value for position Xo by combining a linear system on adjacent data.

λ_i are krigging parameters calculated from the variogram model. Generally the weighting is reduced as the distance approaches the range of the variogram, so krigging uses the variogram to understand the variability in distance.

$Z(x_i)$ These are known values, for example those that exist in the well as a product of scale

---

27 Zakrevsky, K. E. (2011). Geological 3d Modelling. HOUTEN The Netherlands: EAGE Publications.

up.

The variation of calculation produced by krigging according to the range as seen in Fig.3.7 with a Range: 1000m and Range: 10000m in a spherical variogram, clearly the continuity increases proportionally to the value of the range of the variogram. A very small range produces small spatial influences producing very isolated effects.

Fig.3.6 Weighting of krigging based on variogram

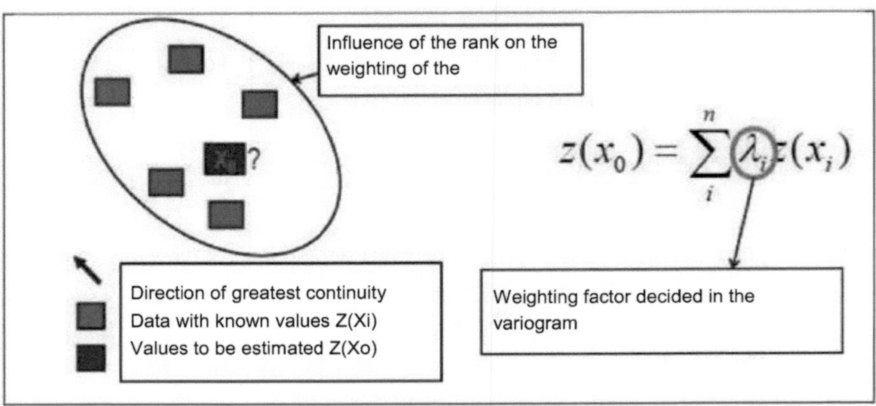

$$z(x_0) = \sum_i^n \lambda_i z(x_i)$$

Source: (Zakrevsky, 2011)

Fig.3.7 Influence of range on krigging

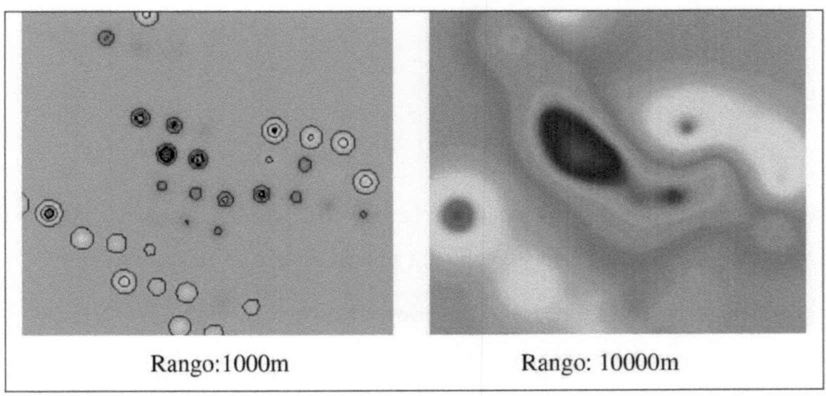

Source: (Dubrule, 2003)

The influence of variogram type (exponential, spherical and Gaussian) on krigging weighting with an areal range of 10000/5000 m as seen in Fig.3.8, in general the results of variogram models give similar results, but with distinctive features.

Fig.3.8 Influence of the variogram model on krigging

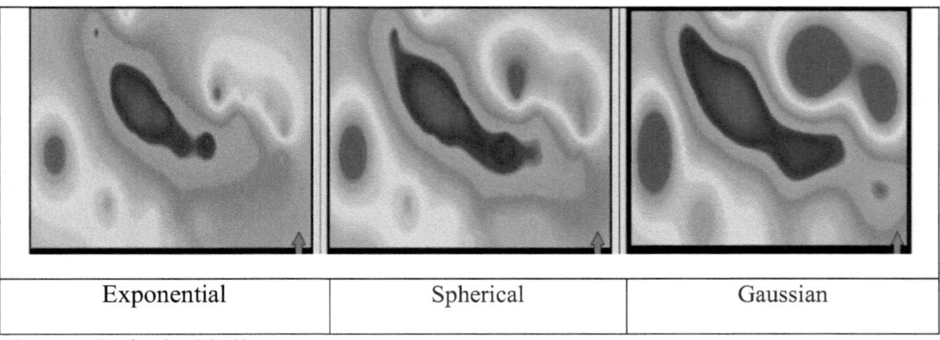

| Exponential | Spherical | Gaussian |

Source: (Dubrule, 2003)

The influence of the nugget for a spherical variogram model as seen in Fig. 3.9 at a value close to 1 will create property maps that look continuous, except at the location of the input data (in the wells) so that the points will acquire values close to the global average in the distances. If you have a nugget value close to zero the results will have much greater variability or heterogeneity within the property.

Fig.3.9 Influence of the nugget on krigging

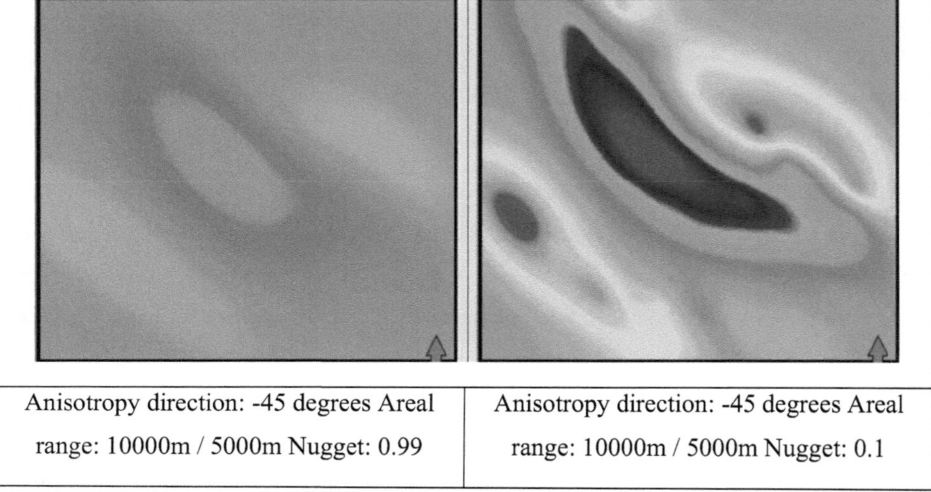

| Anisotropy direction: -45 degrees Areal range: 10000m / 5000m Nugget: 0.99 | Anisotropy direction: -45 degrees Areal range: 10000m / 5000m Nugget: 0.1 |

Source: (Dubrule, 2003)

3.2.3.   Cokriging

Cokriging is a multivariate extension of kriging developed by Matheron in the 1960s, the objective is to predict a rock property at a defined spatial location by combining measurements of the same property at another location with measurements of another variable, this assumes that both variables have a determinable correlation factor, thus seeking to improve the estimation of one variable by a second variable, it is particularly useful when the main variable is of low sample density, compared to the second variable that could be derived from seismic as seismic attributes. Its predictions are better in the sense that it minimizes the estimation of mean square errors. In this sense, a representation of extreme values is achieved, which is important when calculating insitu volumes, or when doing a flow simulation.[28]

3.2.3.   Sequential Indicator Simulation (SIS)

This method is the most popular and is based on the use of the variogram as an indicator and requires well data as input and well calibration, if the different lithologies are identified with discrete values the next step is to define indicator variables to characterize the presence or absence of the facies of interest in the wells. The variogram analysis should not exceed meters for the vertical range, while for the horizontal it should acquire values in kilometers, of course, there is a direct relationship between the proportion of facies of interest and the sill of the variogram.[29]

This modeling method uses the krigging basis but has the advantage of reproducing histograms and complying with spatial variability, which makes it more appropriate for later flow simulations, offering alternative realizations.

This method is a stochastic algorithm, which uses cells with values in the wells to model the facies, the variogram is the parameter used to define the distribution and connection of each type of facies, is a widely used method when you have facies with shapes that are not very clearly defined, or when you have little input data available. Its application may occur at an early stage of field development, when there is no definite knowledge of facies architecture,

28 Coburn, T., Yarus, J., & Chambers, L. (2006). Geostatistics and stochastic modeling: Bridging into the 21st century. In T. Coburn, J. Yarus, & L. Chambers, Stochastic modeling and geostatistics: Principles, methods, and case studies (Vol. Computer Applications in Geology 5, pp. 3- 9). AAPG American Association of Petroleum Geologists.
29 Al-Anezi, K., Kumar, S., & Eibad, A. (2013). Geostatistical Modeling with Seismic Characterization of Wara/Burgan sands-Minagish Field-West Kuwait. Abu Dhabi UAE: Society of Petroleum Engineers SPE 166046.

dimensions, or shapes. SIS can be used to generate preliminary facies models using an input distribution such as a histogram.

The SIS calculation procedure has the following sequence, Fig.3.10:

•   Each grid cell (X3) is visited sequentially following a random path (defined by a start or root number) following a Random function pattern.

•   For each new cell, the local probability distribution function of the property is calculated using the well data and the cells that were previously calculated as control points (X0) (X1) (X2).

•   An estimate of the probability of ownership is estimated using indicator krigging.

•   The estimate is given by the variogram, which is then weighted with the well values.

•   The value calculated for the facies is assigned from the property distribution curve.

•   Finally, the calculated values will be used for the calculation of the next cell by updating the probability curve.

Fig.3.10   Calculation procedure in SIS

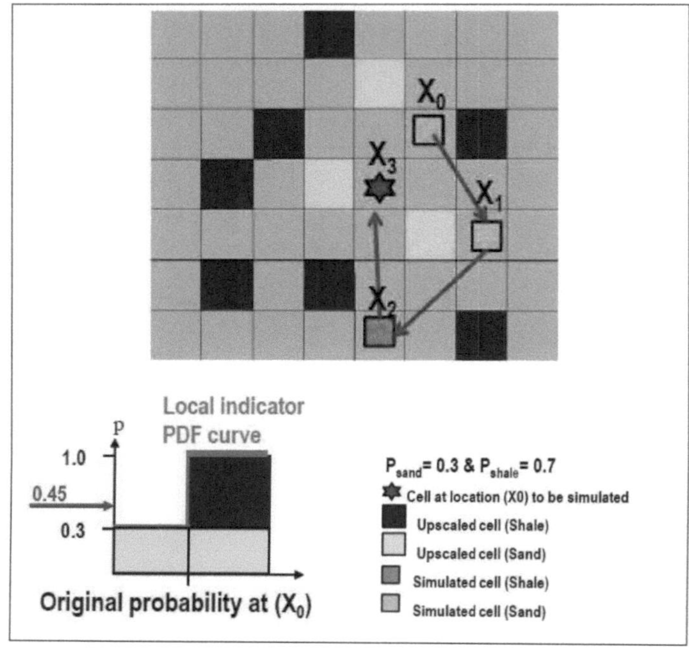

Source: (Zakrevsky, 2011)

The direct influence of the variogram range on the SIS calculation can be seen in Fig. 3.11 where a comparison of small range with undefined anisotropic direction and large range with anisotropic direction with directional tendency is made, clearly these factors influence the property distribution in a preponderant way.

Fig.3.11 Influence of variogram range on SIS

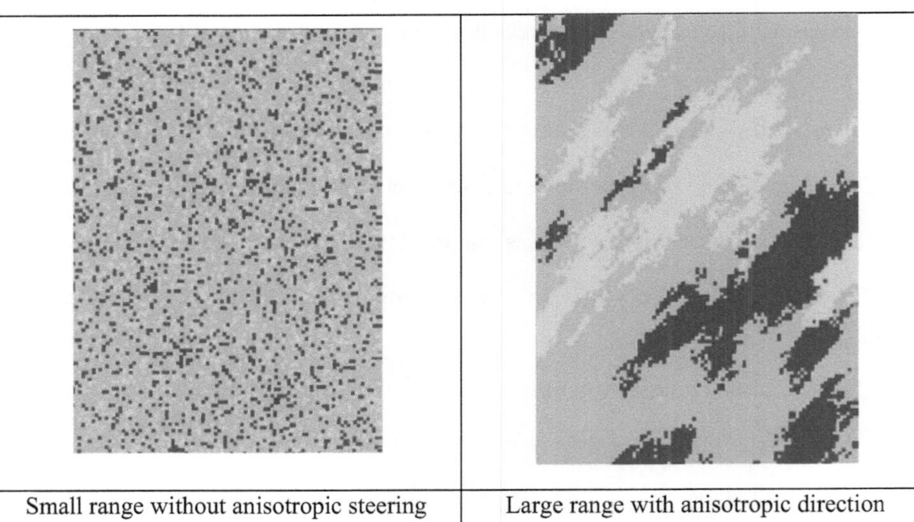

| Small range without anisotropic steering | Large range with anisotropic direction |

Source: (Al-Anezi, Kumar, & Eibad, 2013).

CHAPTER IV

PRACTICAL APPLICATION

4.1. Location and geology of the study field

The field of study for the present work is located in block 15/20a in the North Sea, belonging to the sector of the United Kingdom northeast of the city of Aberdeen, in its surroundings there are neighboring fields 4km east and 15km northeast. It was discovered in 1986 but remained undeveloped until 2010, because it is a low relief structure and its thin oil column of approximately 45ft made this field a marginal prospect with little economic attractiveness. A critical factor for its development was the installation of the production facilities which presented challenges in terms of design and installation. This field was discovered through well 15/20a-3 which was drilled to evaluate the Paleocene potential in a four-way fault system that forms a structural closure, this well encountered a 75ft column of oil in turbidites interbedded in the Andrew formation with structural location in the northern part of the graben.[30]

The seismic data used for the interpretation were acquired in 2005 for the purpose of field development. In 2008 this volume was reprocessed in an attempt to improve the quality of the signal and reduce the presence of noise, for this a preSTM (pre stack time migration) processing was used. Along with this work was also requested differentiated processing to generate seismic cubes based on reflection angles in nearstack (7-15), midstack (15-30), farstack (30-45), ultrafarstack (45-60) and a compiled in fullstack, this type of processing was performed in order to perform studies of amplitude anomalies (AVO Amplitude versus offset) in post stack and in correlation with the reflection angles.

The AVO studies at Prestack were conducted internally at the operating company and their publication and disclosure is confidential.

The present project did not have access to constructed or interpreted surfaces, petrophysical evaluation or 3D grid constrained models, only to raw data, i.e. seismic cubes in poststack, fullstack, near stack, far stack, conventional electrical records (Caliper, Bulk Density, Density correction, Compressional slowness, Gamma Ray, Neutron porosity, Photoelectric

---

30 Gluyas, J. G., & Hichens, H. M. (2003). United Kingdom Oil and Gas Fields (Vol. Commemorative Millennium Volume). London, UK: GEOLOGICAL SOCIETY.

factor, Deep resistivity, Medium Induction, Micro Resistivity, Shallow Resistivity, Shear Wave Slowness, Potassium Concentration, Spontaneous Potential, Thorium Concentration, Uranium Concentration), deviation logs (MD, Dip, Azimuth), and checkshots for three wells.

The arrangement of wells as well as the area of coverage of the seismic record can be seen in Fig.4.1, where there are 15 wells that were used in the present study.

Fig.4.1 Spatial arrangement of borehole and seismic data

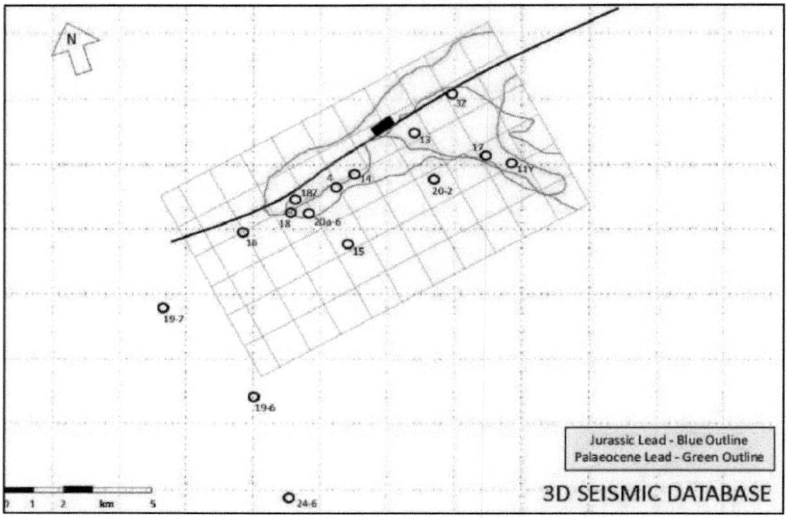

Source: Geological report

Within the regional geological information, it can be found that this structure is characterized as an extensional basin with predominant northwest-southeast structural direction, having contact in the northern part with the basement high known as Fladen Ground Spur, the extensional period occurred between the Triassic and Cretaceous, expressed in normal faults of listric type, the main phases of faulting control occurred in the Lower Cretaceous generating low-relief structures as shown in Fig.4.2 which highlights the preponderant geological time in the area of interest, above these faults is the productive interval corresponding to sands from turbidic flow of a variety of characteristics.

Fig.4.2 Geological identification of faulting

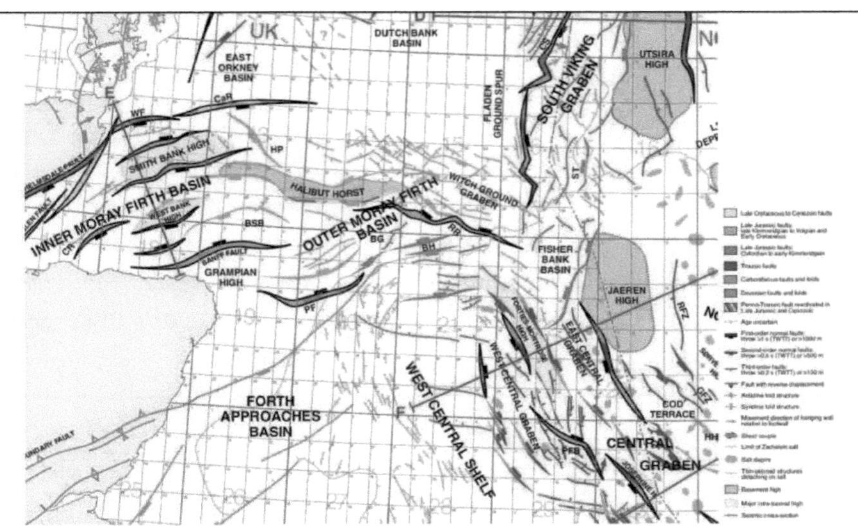

Source: (Gluyas & Hichens, 2003)

Regarding the petroleum system, within the North Sea study area, the anoxic sediments belonging to the Upper Jurassic known as Kimmeridge Clay Group are shales with a high organic content, have upper contact with the Early Cretaceous Cromer Knoll Group, this formation is known to be the main source rock in the area, and where most of the hydrocarbon production comes from. The sedimentation process is controlled by the interaction of Eustatic/Tectonic processes, referred to as transgressive and regressive processes, which are divided into three main parts. The main transgressive pulse was the cause for the deposition of the Kimmeridge clay formation, this event initiated changes in the chemistry of the water bodies and a lithological change, but without major changes in the sedimentation process within the isolated basins in their communication with the open ocean; that is, this source rock was probably deposited in a restricted sea space, with low oxygen levels in the basin centers and with restricted water flow circulation. At present, this source rock is sufficiently mature throughout the area, and even in deep places it becomes over-mature and reaches the gas generation stage. Immediately after the deposition of the reservoir sands, the source rock was mature in the locality near the basin depocenters. Table 4.1 summarizes the properties and formations identified in the petroleum system.

Table.4.1 Oil system summary

Seal	rockSele Formation(Clay) Lista formation

Paleocene	(Clay)	
Reservoir rock Paleocene	Upper Balmoral Member (sand) Forties formation (sand) Maureen formation (sand) Thor formation(chalk)	Good porosity and permeability according to 15/24b-6 well DST
Source rock Jurassic	Kimmeridge formation (Clay)	Thermal maturity (Ro= 11.15) near lower oil floor. Hydrocarbon generation potential (TOC=7-10%) good excellent

Source: Geological report

4.2. Drilling reports

The total number of wells drilled in the field is 21 in both the exploration and development phases, of which 17 are development wells and the remaining are exploration wells. The electrical logs acquired are conventional, among which gamma ray, resistivity, density and neutron were acquired in all wells. Formation pressure tests are present in 6 wells and only 3 have production DST tests, Table 4.2 illustrates the type of logs that each well has and which were accessed for this study.

Table.4.2 Electrical logging information available per well

REGISTROS ELECTRICOS			POZOS														
			15_20a-3Z	15_20a-4	15_20a-6	15_20b-11Y	15_20a-13	15_20a-14	15_20b-15	15_20b-16	15_20a-17	15_20b-18	15_20b-18Z	15_19-6	15_19-7	15_20-2	15_24b-6
CALI	.in	: Caliper															
DENS	.g/cc	: Bulk Density															
DENSC	.g/cc	: Density correction															
DTC	.us/ft	: Compresional slowness															
GR	.gapi	: Gamma Ray Log															
NEUT	.frac	: Neutron porosity															
PEF	.b/e	: Photoelectric factor															
RDEP	.ohmm	: Deep resistivity															
RMED	.ohmm	: Medium Induction															
RMIC	.ohmm	:Micro Resistivity															
RSHAL	.ohmm	:Shallow Resistivity															
DTS	.us/ft	:Shear Wave Slowness															
POTA	.%	:Potassium Concentration															
SP	.mV	:Spontaneous Potential															
THOR	.ppm	:Thorium Concentration															
URAN	.ppm	:Uranium Concentration															

Source: Geological reports

4.3. Stratigraphy

The general stratigraphy of the reservoir rock corresponds to the Balmoral lower member belonging to the Lista formation within the Moray and Montrose groups, in the upper Paleocene, Paleogene in the Tertiary as shown in detail in Fig.4.3. This rock is a sandstone is characterized by a heterolitic lithology of intercalation of shales and sands, is subdivided into 5 zones that are genetically related, and have sandy trend defined as M2-M6, these zones are separated by biostratigraphic variations where the shale trend dominates defined as S1-S5, this classification is made from analysis studies of cores extracted in the wells 15_20a-4, 15_20a-6. The top of the reservoir is represented by the base of the S5 interval. To the north the reservoir sands are M4b, while M5 sands are reservoir in the southern part. This stratigraphic framework is further complicated in some localities by the presence of sand injectites, as shown in Fig. 4.4, which also shows the expression of the zones identified in gamma ray and their typical lithology markers.

The depositional types found in the formation correspond to different types of flows, among them turbidites of high and low density, mixed with hemipelagic deposits; injectites also appear, this produces a variation in the types of lithofacies. The equivalence in lithofacies and their typical scale can be observed in Table 4.3.

Fig.4.5 shows the base correlation of the sand M4a, and M4b where the oscillation of the sedimentary bodies is seen across the structure, zones of thin sand and shale with northeast direction for M4b with a decrease in thickness towards the flanks, this correlation is across the structure axis, indicating that M4a and M4b form offset depositional intervals at the boundaries.

Fig.4.3 General stratigraphy

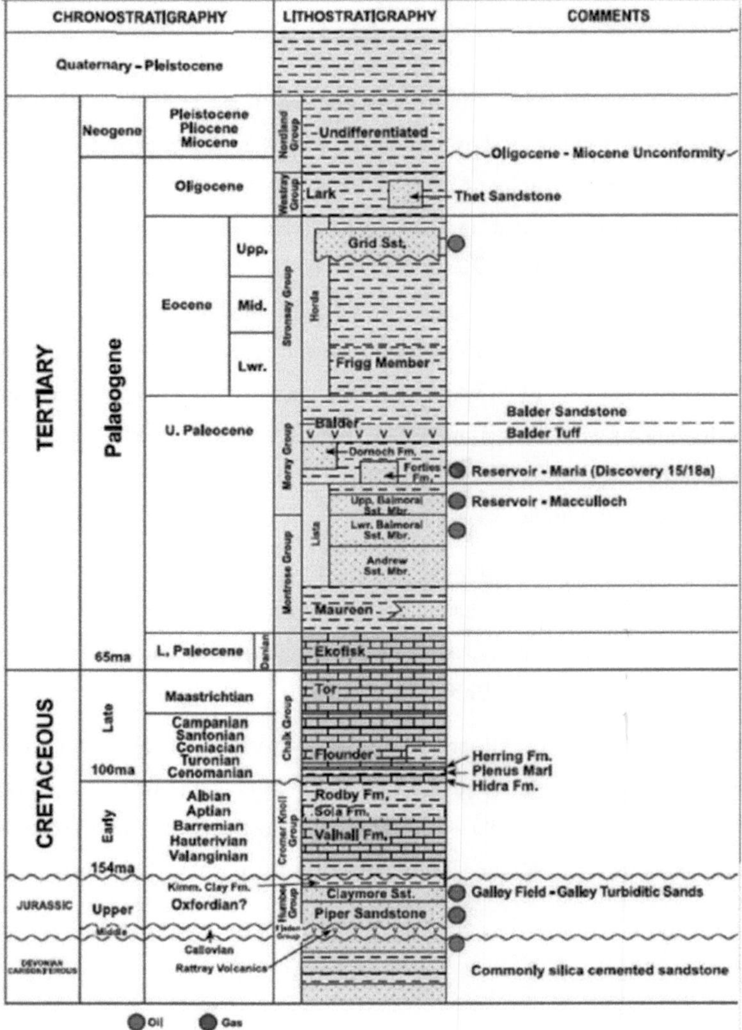

Source: Field report

Fig.4.4 Stratigraphic classification and expression in records

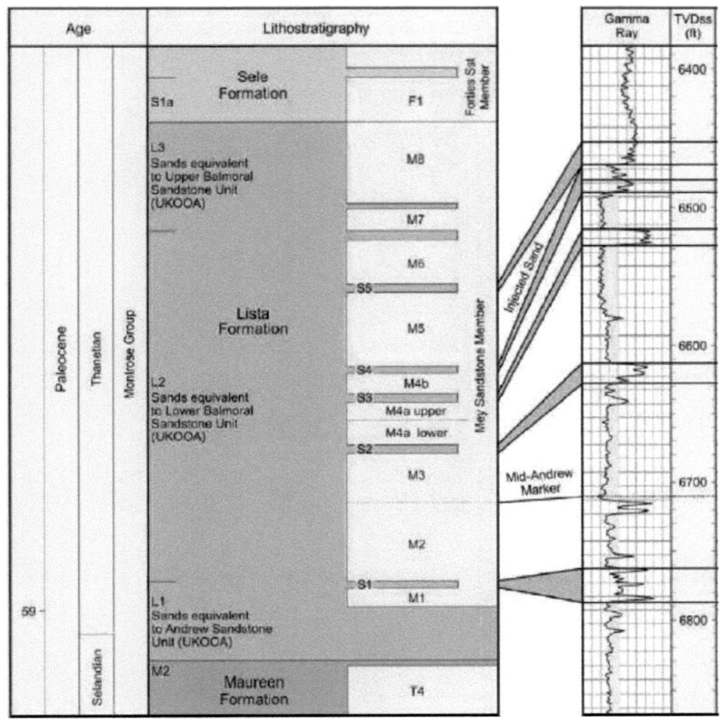

Source: (Gluyas & Hichens, 2003)

Table.4.3 Depositional type and lithofacies description

Depositional process	Lithofacies	Scale (ft)
High-density turbidite	Amalgamated massive sandstone	1-30
	Amalgamated massive sandstone with granules Layered massive sandstone	1-15 <1
Slurry flow	Banded sandstone	1-15
Cogenetic turbidite-debrite	Clast-rich muddy sandstone 'linked debrites'.	0.5-7
Low-density turbidite	Thin-bedded heterolithic strata	0.5-30
Hemipelagic	Burrowed claystone	0.5-30
Injectite	Small discordant sandstone dykes	<1
	Injection breccia	0.5-3
	Sandstone injections > 1 ft Remobilized massive sandstone	1-6
		0.5-7|

Source: (Gluyas & Hichens, 2003)

Fig.4.5 Lithological correlation for M4a and M4b sands.

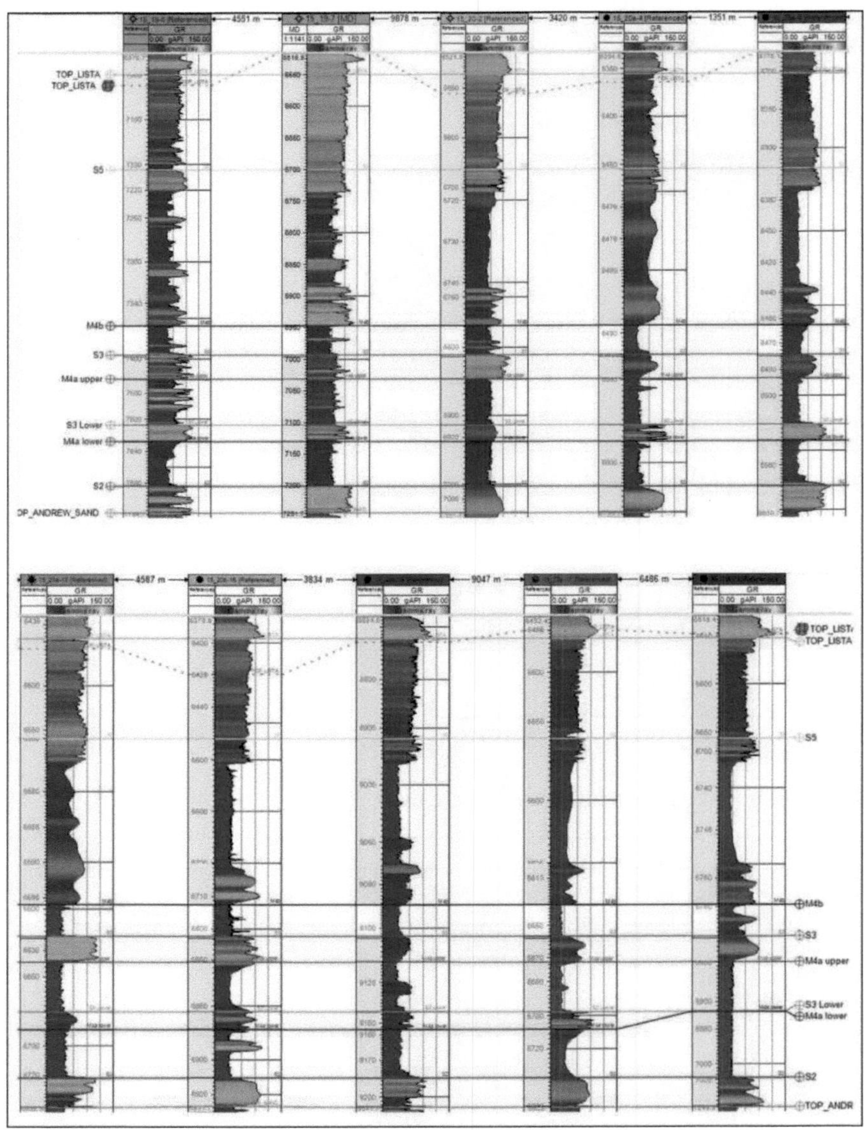

Source: Own elaboration

Fig.4.6 shows the basic evaluation of electric logs for wells 15/19- 6, where the objective is the integration of data such as: formation markers, reference chronostratigraphy, as well as the neutron density crossover to highlight the effects of gas. According to the drilling reports, only traces of hydrocarbons were found in this interval, and in caliper breakouts can

44

be seen in the interval making the readings in these zones unreliable, high resistivity readings can also be seen, but in a very laminated way and some do not match with neutron density. Despite the fact that this well came out dry and only water was encountered, it shows very good indicators of the presence of hydrocarbons.

Fig.4.7 shows the electric log data integration process for well 15/19- 7, being that the caliper readings show breakouts in the upper part, makes it difficult to recognize sandy members in this interval which could affect the resistivity readings, as in the previous case this well found hydrocarbons in the Jurassic, but not in quantities that can be considered as potential producers.

Fig.4.8 shows the data integration for well 15/20- 2 where the decrease of the neutron density gas effect can be appreciated and where the resistivity peaks are much more laminated and punctual, the caliper readings make the sand intervals not completely reliable in their recognition.

Data integration shown in Fig.4.9 for well 15_24b- 6, where a gamma ray interval that does not show much variation between shale and sand can be seen, however a high level in resistivity can be seen, quite continuous although oscillating where the productive interval is located, this is corroborated by the neutron density crossover, belonging to the Paleocene, this well in the DST production tests gave a flow rate of 3600 Bbld of production.

Fig.4.6 Data integration for well 15_19-6

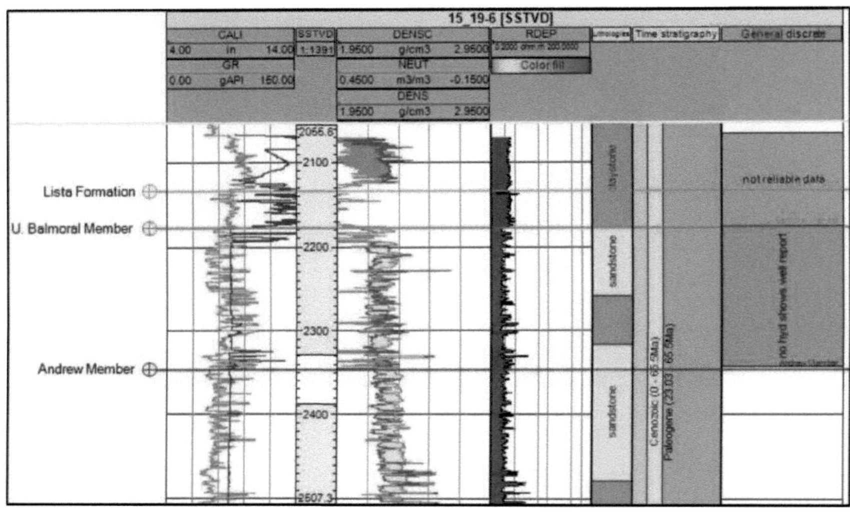

Source: Own elaboration

Fig.4.7 Data integration for well 15_19-7

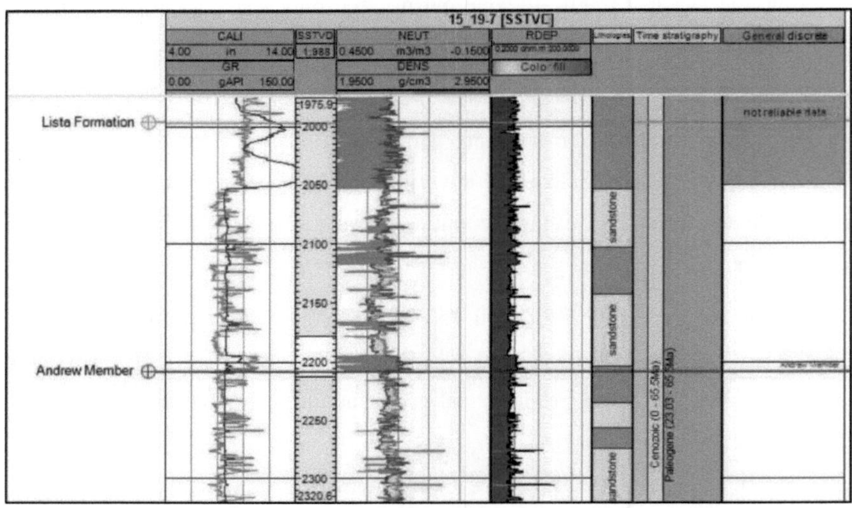

Source: Own elaboration

Fig.4.8 Data integration for well 15_20-2

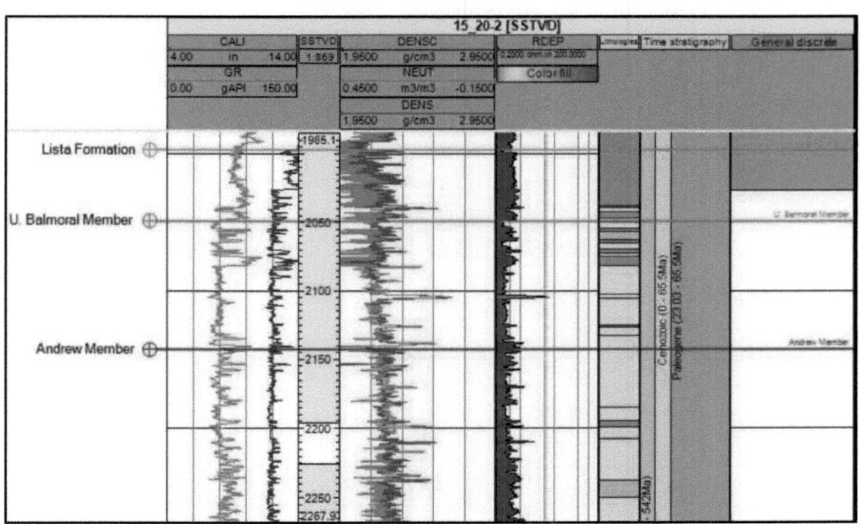

Source: Own elaboration

Fig.4.9 Data integration for well 15_24b- 6

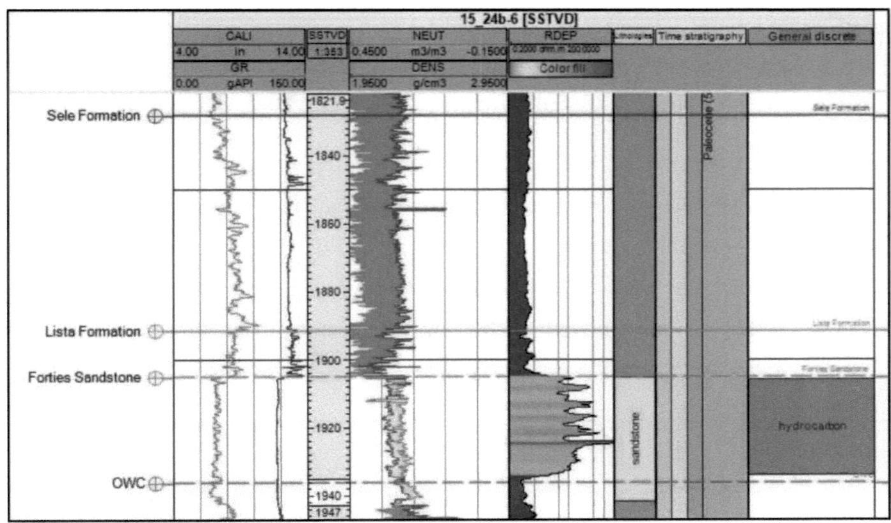

Source: Own elaboration

The facies classification was made considering a very important aspect, referred to the reactivity of the sands, with a high content of heavy minerals (potassium, thorium) that make the gamma ray readings much higher in sand intervals, this makes it necessary to modify the weighting of this record when classifying facies, where higher gamma ray readings also correspond to reactive sands as shown in Fig.4.10.

Fig.4.10. Identification of reactive sands in well 15_24b- 6

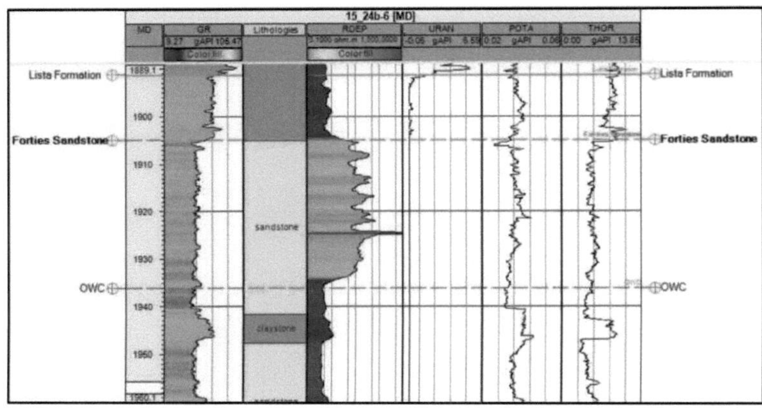

Source: Own elaboration

Fig.4.11 shows the continuity of the identified formations, showing their variation in thickness and continuity, as well as the variations in depositional characteristics, the log used for this comparison is gamma ray and the variation between massive sand and shale bodies, the appearance of sandy clay intercalations and absence in other cases can be seen.

Based on the above criteria, facies classification was performed using gamma ray, density, sonic and neutron logs for four predominant facies such as Sand, Fine Sand, Coarse Sand and Shale: Sand, Fine Sand, Coarse Sand and Shale, and obtain the stratigraphic column closest to the lithological behavior in these logs, the result can be seen in Fig.4.12 where clearly the transition from massive to more laminated and condensed bodies can be seen.

Fig.4.11. Correlation of lithology

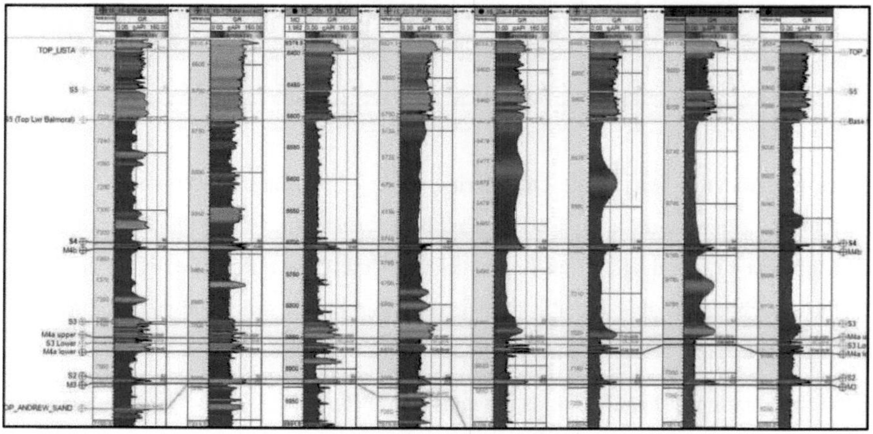

Source: Own elaboration

Fig.4.12 Interpretation of lithological facies

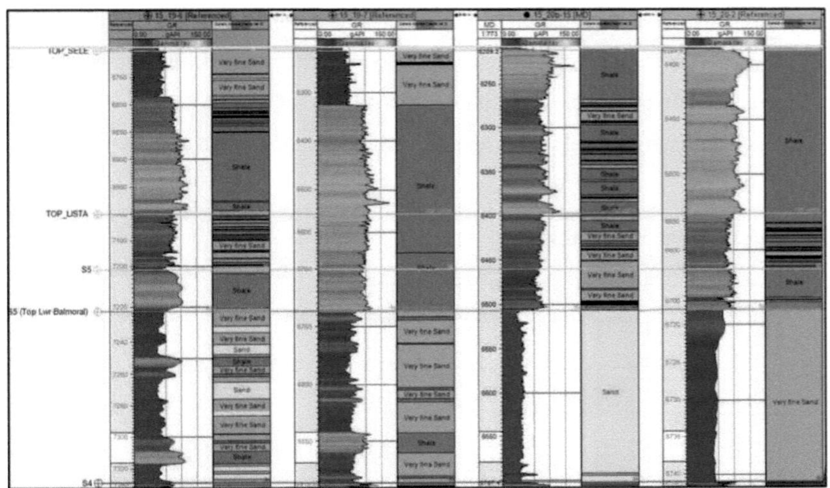

Source: Own elaboration

4.4. Analysis of seismic expressions

Within the structural aspect the field has sub-seismic faults and fault boundaries that were recognized in the drilled wells. In the seismic data it is possible to appreciate intervals with southwest inclination in the axis of the basin, where the sands that can be visualized are massive of laminar type.

For seismic interpretation, identifying and tracking the reservoir top reflector is difficult because of polarity reversal in full stack processing, due to the presence of oil and fluid filled sands, so the use of color manipulation, signal and structural attributes are required to achieve a proper interpretation.

The importance of seismic expressions is given not only at the time of seismic interpretation, but also to correlate reflector patterns with depositional significance, in this sense the following reflector expression convention is used as a basis for characterization, as shown in Fig.4.13:

- Parallel or subparallel, wavy, divergent, in general.

- Sigmoidal, complex sigmoidal or oblique, parallel oblique, stretched, tangential oblique, hummocky type, in which sedimentological factors are expressed.

- Discontinuous, contorted, lenticular in those of continuity.

The configuration of the reflectors depends on the shape and angle at which sedimentation occurs and is influenced by: the composition of the deposited material, the sedimentation flow rate and the amount of sediment available, the salinity of the water, the depth with respect to sea level, the energy level of the depositional environment, the subsidence radius, among others.[31]

Fig.4.13 Classical sedimentary expressions

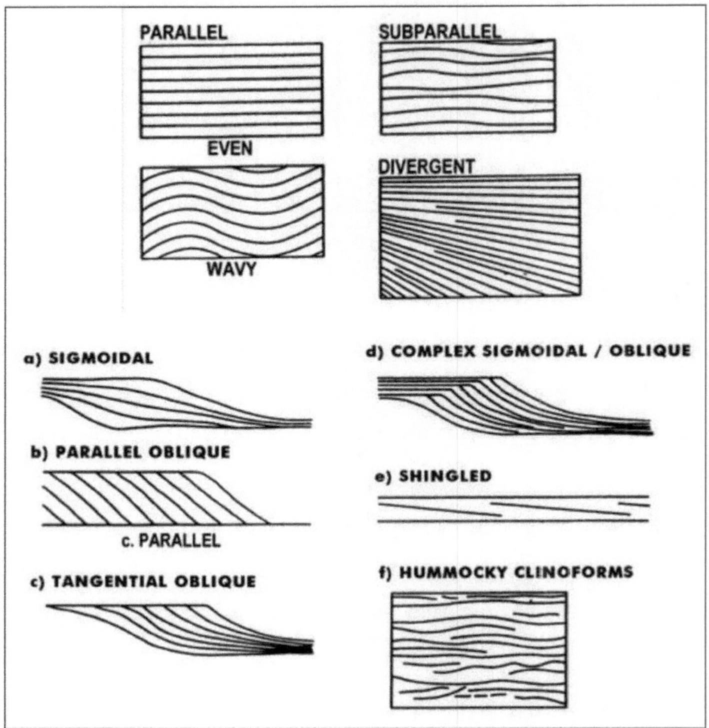

Source: (Veeken & Van Moerkerken, 2013).

The seismic expression in the Fullstack seismic cube for the sedimentary bodies can be seen in Fig.4.14 where the interpreted reflectors are converted horizontally or reconstructed to highlight the depositional character, it can be clearly seen that on the left side the expressions are subparallel in a mix between oblique parallel and hummocky clinoform lenticular and contorted, this behavior is typical of mass transport flows. On the right one

31 McKie, T., Rose, P., Hartley, A., Jones, D., & Armstrong, T. (2015). Tertiary Deep-Marine Reservoirs of the North Sea Region (Vols. Special Publications, 403). London: Geological Society.

can see more continuous disrupted but contorted reflectors.

Fig.4.14 Seismic expressions in a reconstructed Xline shear

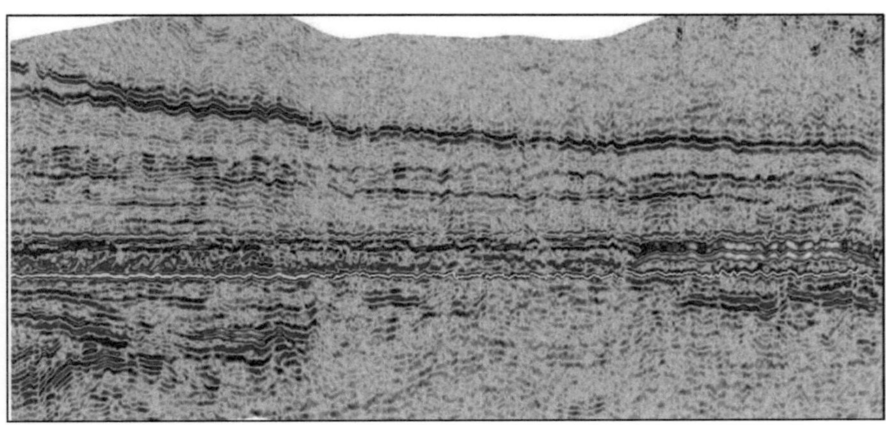

Source: Own elaboration

For the interpretation of base horizons, the markers of well 15_24b-6 were defined as a starting point, where there are high faulting characteristics and where the reflectors are more continuous and identifiable in the upper part while in the lower part the reflectors are less continuous and more of a disrupted type, between interpreted reflectors lenticular shapes typical of depositional flows are observed as shown in Fig.4.15.

Fig.4.15 Seismic interpretation with faults in borehole 15_24b-6

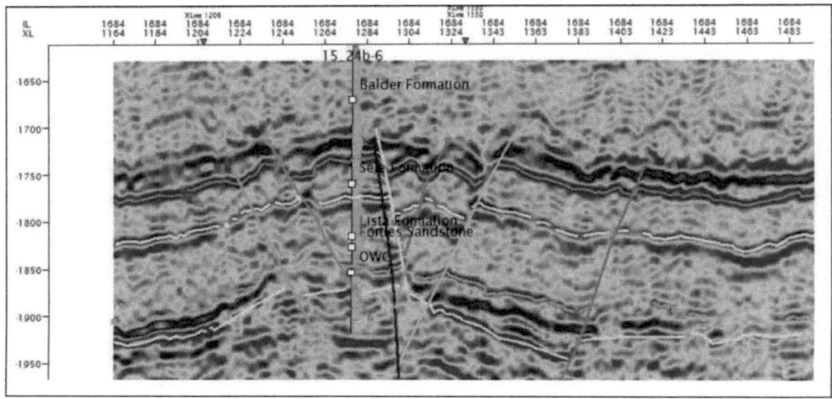

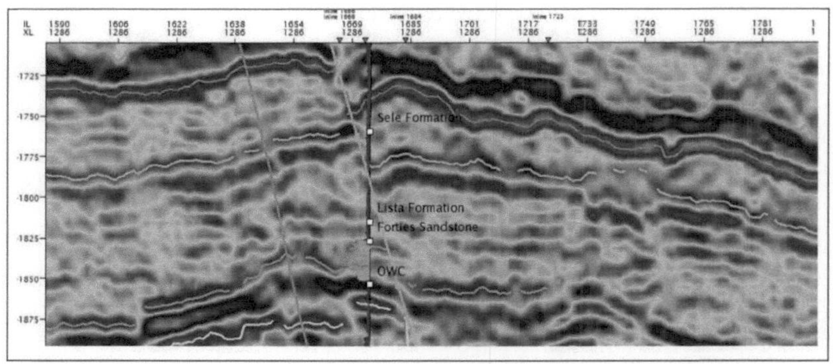

Source: Own elaboration

Among the seismic attributes used to identify sedimentary bodies are Coherence as seen in Fig.4.16 where the depositional channels of the turbidites are clearly visible with different directions, the continuity of the bodies is modified due to the inclination of the bodies. One can also see changes in flow direction by order of age of deposition.

Fig.4.16. Coherence attribute for sedimentary bodies.

Source: Own elaboration

Among the most important features that can be observed in two seismic data are those that express gas columns as shown in Fig.4.17, evidencing the existence of a current petroleum system and the existence of fluids that have a migratory process towards porous bodies, product of this can be seen the increase of the amplitude of some reflectors making them brighter, which implies a variation of the acoustic impedance contrast.

Fig.4.17 Seismic gas columns

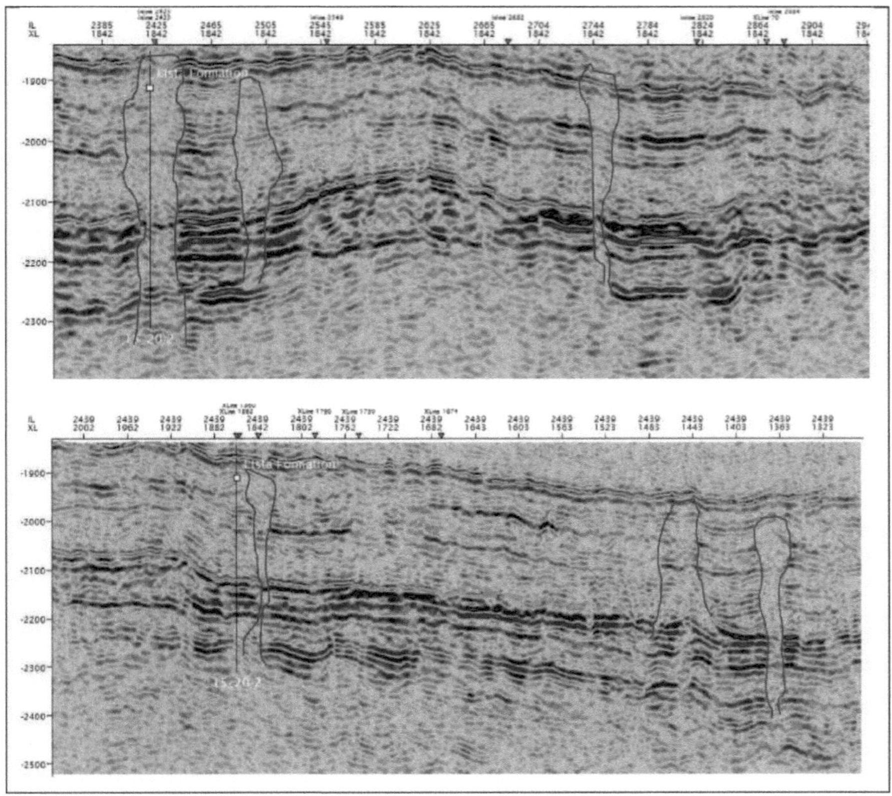

Source: Own elaboration

4.5. Geological model

The field has formed a closed structure in four directions as shown in Fig.4.18, the seal rocks are shales of the Lista and Sele formations, the reservoir rock accumulation is complete in the western part with hydrocarbon columns up to 100 ft above the aquifer which is at least 700 times larger in volume. Analysis of the petroleum system shows that the source rock corresponds to shales of the Jurassic Kimmeridge formation and that the fluid has migrated laterally and vertically into the trap, where one of the factors that contributed to the migration are the sand injectites in the clayey part that created permeable

53

flow channels, ideal for migration and accumulation.

Fig.4.18   Structural map of the field of study

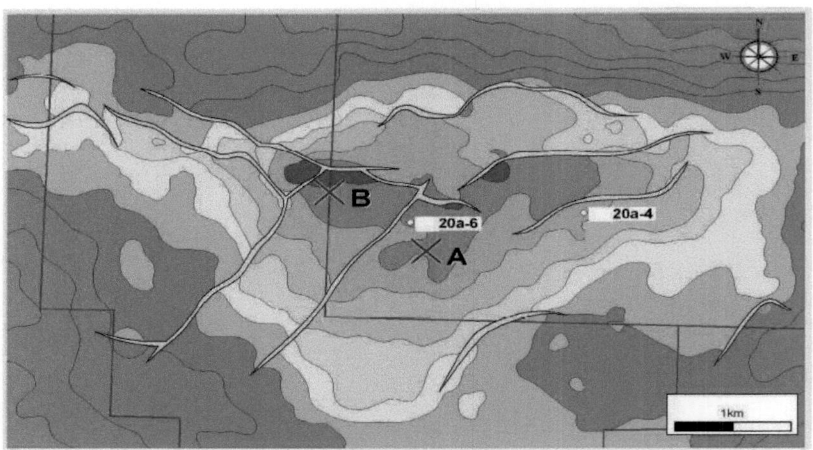

Source: Field report

The field is located above cortical faults that reach the basement in a northwest-southeast direction, these faults were active during the Late Jurassic to Early Cretaceous, with episodes of reactivation during the Tertiary. Structural stability occurred from the Late Cretaceous to the Early Paleocene. However, minor movements on these faults probably influenced sand deposition in the Balmoral Formation. Reactivation and inversion occurred during deposition of the Balder Formation and during the Miocene. Some faults that were originally extensional during the Jurassic became transpressional faults during late Alpine orogenesis.

The regional area where the field is located is dominated by amalgamated sandstones, hemipelagic clays, sandy injectites and clay-prone sands, these characteristics are the most significant heterogeneities.

The identification of injectites in sands identified in the cores is an important factor in explaining the migration behavior and flow dynamics of the reservoir, as well as the juxtaposition of different zones through small-scale faults which are more common in the northern area.

The Lower Balmoral formation sands have porosities above 25% and permeabilities greater than 800mD, the reduction of these properties is due to very fine marginal grains, and the result of a more distal position of the flow system location. Reservoir quality is mainly controlled by texture, grain size variation and sorting. The best reservoir quality is observed within the clean, massive sands that are dominant in laminated amalgamated channels where permeability is greater than 1000mD with net to gross radius greater than 90%. In the distal lobes the deposits are composed of thin debritic sediments, with a high clay content in the matrix.

The 3D geologic model used for field development was simplified for reasons of structural representation, seismic quality improved fault resolution and structural consistency. Structural interpretation and recognition of subtle structural features increased uncertainty, making it necessary to consider alternative geologic scenarios.

Fig.4.19 shows the structural top of the Sele formation, where a high level of faulting can be evidenced, which is accentuated due to the lithological characteristic, shales with a high level of dewatering characteristics, which generates the rhomboidal and erratic geometry, the delineation of faults has a polygonal shape and becomes much more perceptible being that the structural analysis for the construction of the model is based on this horizon.

The final geological model includes the structural tops of the formations as shown in Fig.4.20, where the modeled faults are those found in the area of interest, being the most outstanding and representative, since they present amplitude anomalies, although they are the main and the most important in terms of displacement and resolution, it is necessary to highlight that there are faults below the seismic resolution that cannot be visualized but their analysis would be very important to characterize the migration and production mechanisms, but their study would lead to another project.

Formation tops are shown in Fig.4.21- Fig.4.22- Fig.4.23- Fig.4.24- Fig.4.25 as well as modeled faults, all of them were represented based on seismic interpretation without the support of FMI logs to serve as correlation between the borehole and available seismic data, the model has a high degree of continuity across the stratigraphic interval of interest, and prioritizes the continuity of the structure over the faulting detail.

A correct evaluation process should have structural analysis data based on FMI log interpretation, where the families of faults within the interval of interest are identified, and contrast this information with the same characteristics in a steronet plot.

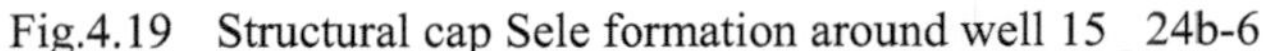

Fig.4.19   Structural cap Sele formation around well 15_ 24b-6

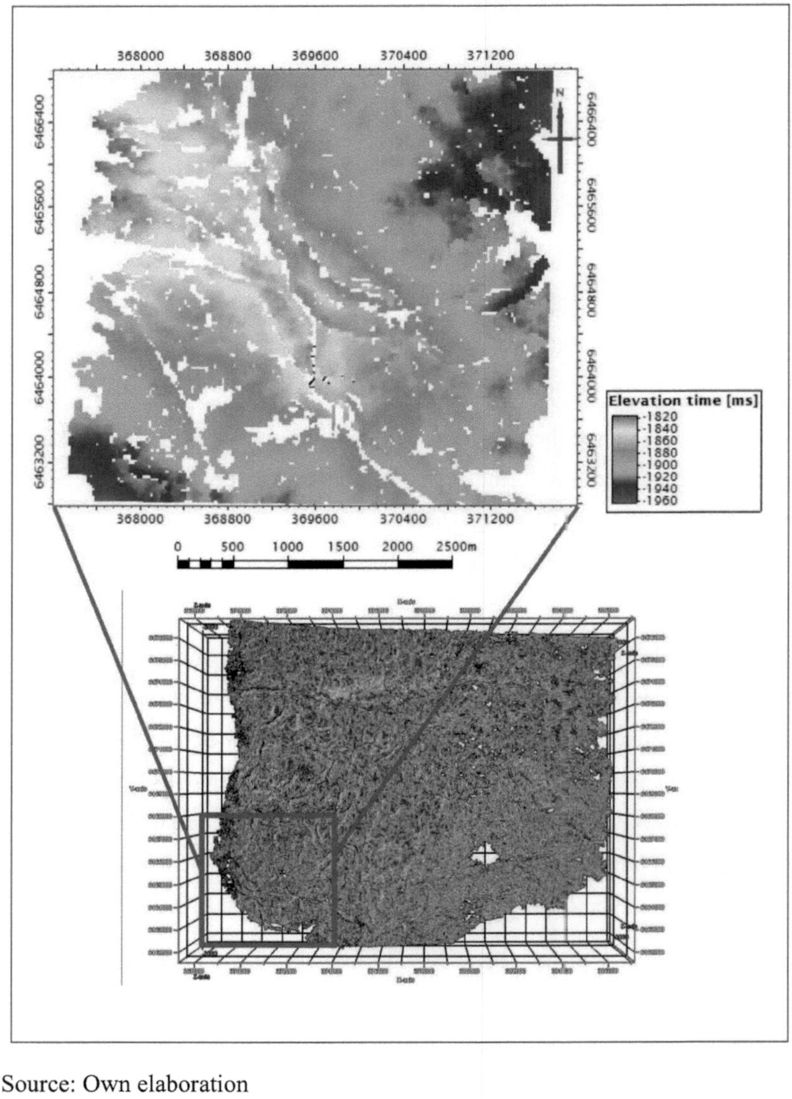

Source: Own elaboration

Fig.4.20 Structural volume with modeled faults

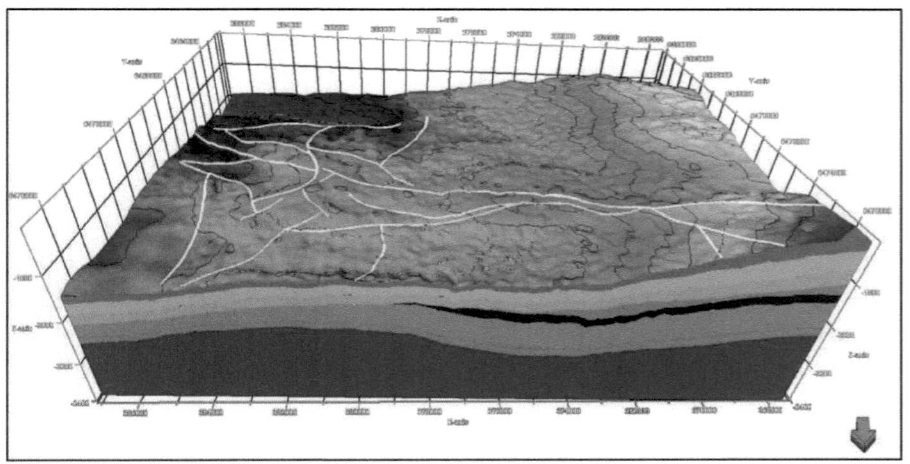

Source: Own elaboration

Fig.4.21 Top of the Balder Formation

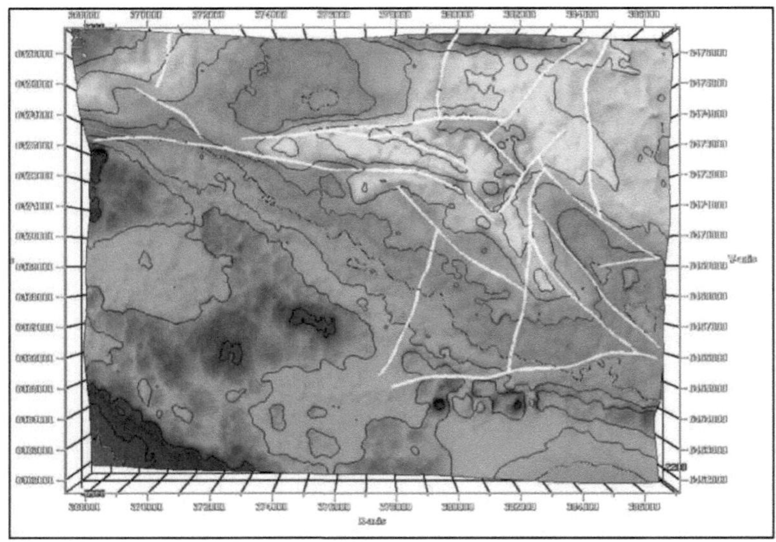

Source: Own elaboration

Fig.4.22 Top of the Sele Formation

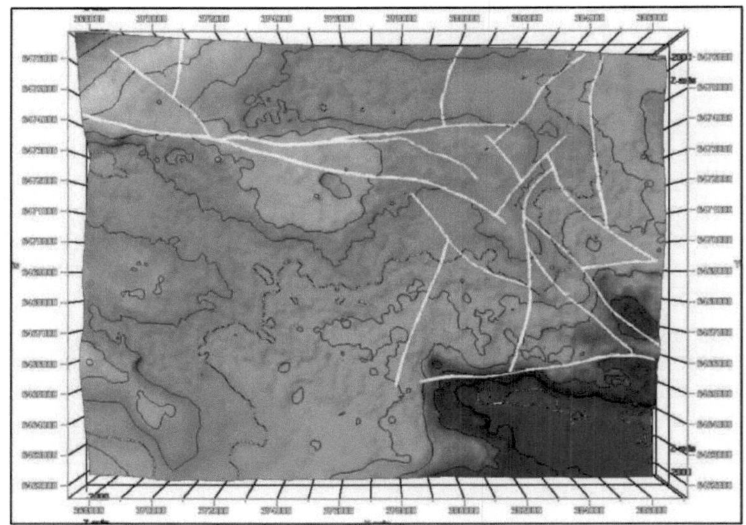

Source: Own elaboration

Fig.4.23 Top of the ready formation

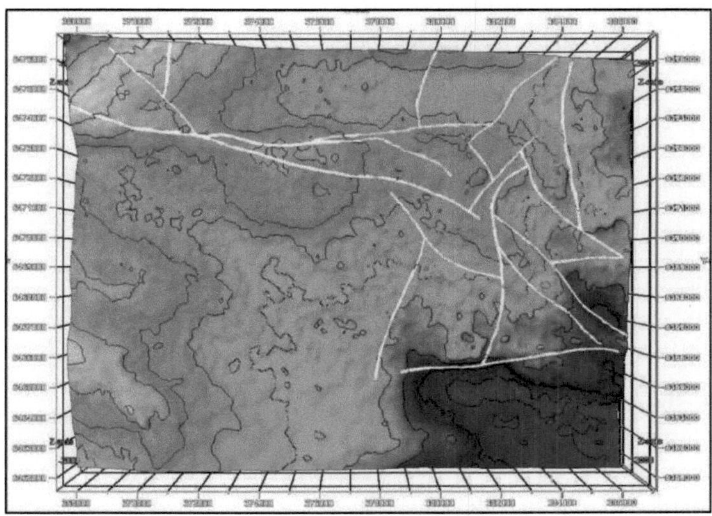

Source: Own elaboration

Fig.4.24 Top of Base member S5 (Top Lwr Balmoral)

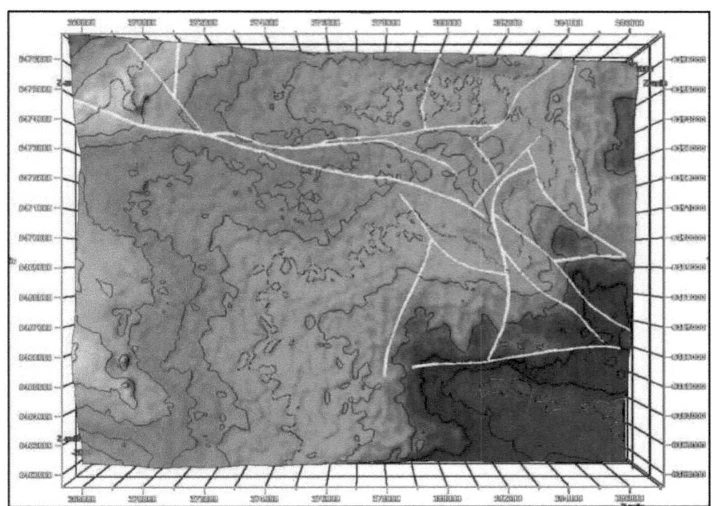

Source: Own elaboration

Fig.4.25 Top of the Maureen Formation

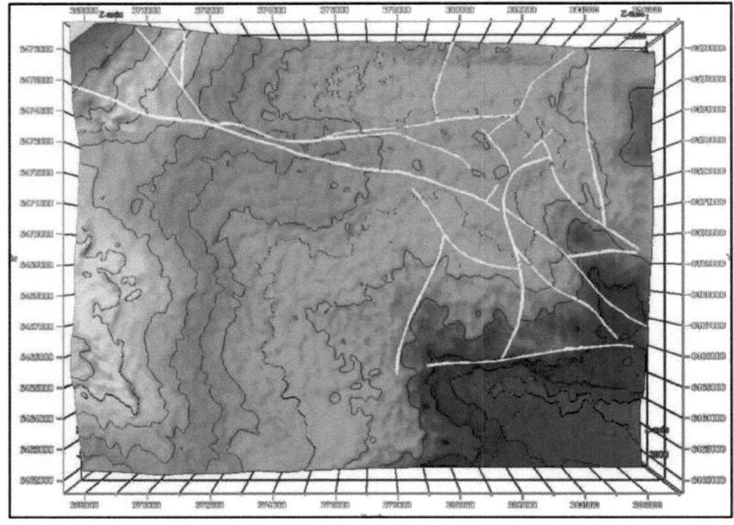

Source: Own elaboration

4.6. Scale up process and vertical variogram analysis

Once the construction of the geological model is finished, the process of transferring well data to the model must be carried out; that is, from electric logs to the grid cells, this process is known as scale up, for its correct implementation it is necessary to take into account the trajectory of the well, the cells that are in its path, as shown in Fig.4.26. The assignment operation that must go in geological consequence with the representation sought and the type of variable to be represented, for facies is of discrete type, it is obvious at this point to emphasize that the smaller the resolution of the cells the more representativeness will be achieved, however the number of cells will increase exponentially requiring the hardware to its maximum capacity.[32]

Fig.4.26 Example of scale up process

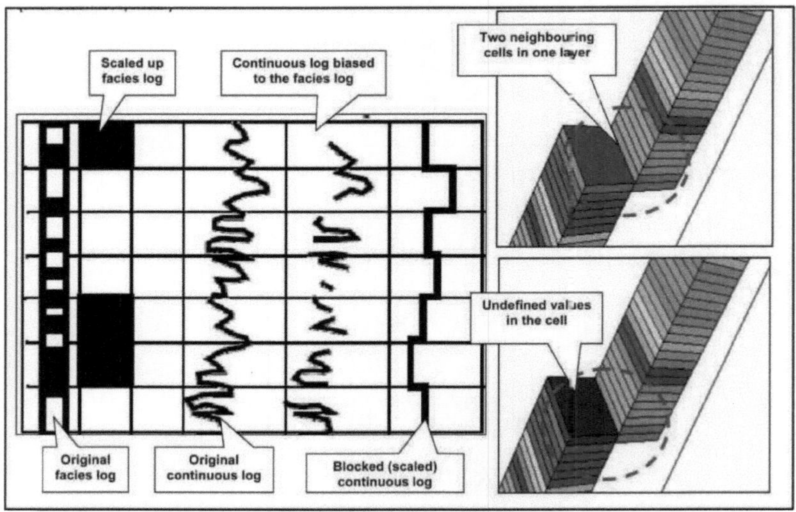

Source: (Zakrevsky, 2011)

For the application in the project, an interpretation of the most important facies within the lithological characteristics such as grain size in the facies of: Sand, Fine Sand, Coarse Sand and Shale. The scale up procedure takes into account these logs and assigns values to the cells according to the number of facies that cross the well trajectory, the algorithm used is of type "Most of" that makes a weighting to assign facies in percentage of presence of the variable. As shown in Fig.4.27 the facies assignment was made for different values of vertical resolution in the grid, track 1 corresponds to gamma ray, track 2 to the facies interpreted in the electric logs, track 3 corresponds to a grid with a vertical resolution of 5 ft, track 4 corresponds to 13 ft, and track 5 to 25 ft resolution respectively, the comparison of

---

32 Zakrevsky, K. E. (2011). Geological 3d Modelling. HOUTEN The Netherlands: EAGE Publications.

representativeness of the facies according to the vertical resolution can be observed.

Fig.4.27 Comparison of interpreted facies assignments in scaleup

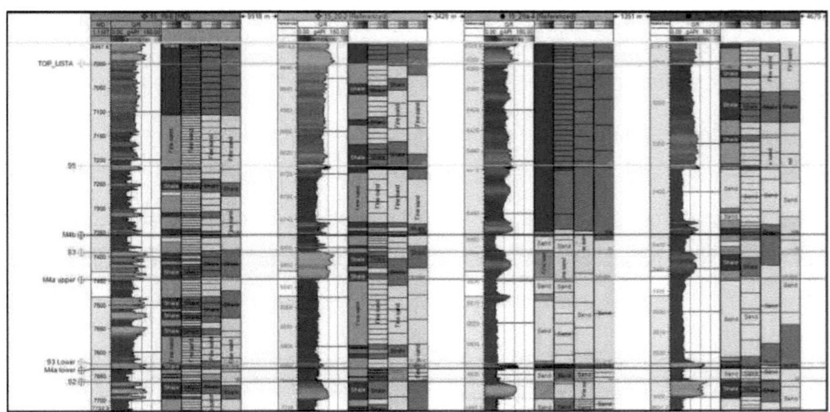

Source: Own elaboration

The experimental variograms were constructed based on the interpreted facies resulting in an Exponential variogram model with a Sill= 0.2606 and a vertical range= 44.008 these values as can be seen in Fig.4.28.

Fig.4.28 Model and experimental vertical variograms for facies

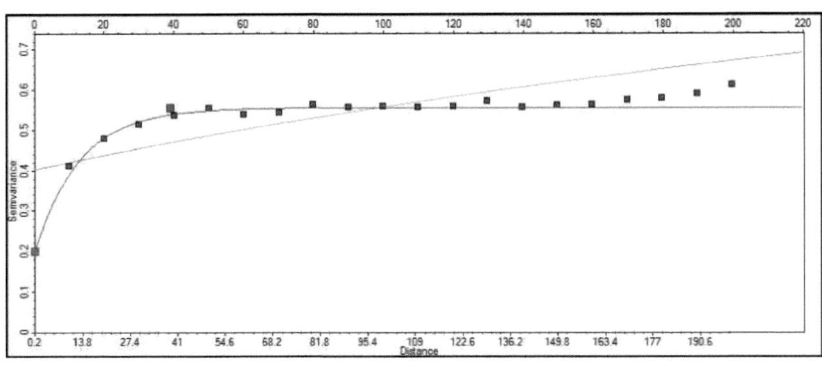

Source: Own elaboration

4.7. Stochastic modeling applying sequential indicator simulation

To make the facies model, we proceeded to recognize the paleoflows in the first instance taking into account the seismic attributes of RMS and Coherence on the interpretation surfaces, thus achieving the identification of dominant sedimentary directions that in the case of TOP_SELE in Fig.4.29 has a north-south trend, for Mb4 northeast-southwest in Fig.4.30 and for TOP_MAUREEN north-south more uniform in Fig.4.31..29 has a north-south trend, for Mb4 northeast-southwest in Fig.4.30, and for TOP_MAUREEN north-south more uniform in Fig.4.31, the directions found in the maps are consistent with the bibliographic referential direction.

The result of the application of the sequential indicator algorithm has a very good distribution and representativeness in terms of continuity and facies intercalation in the study interval, as seen in Fig.4.32 for S3 to the top of the Thor formation using the experimental variograms shown above with variables subjugated to the RMS and Coherence attribute as cokriging, being the distributions of Sand=36%, Fine Sand=18%, Coarse Sand =10%, Shale = 36%.

Fig.4.29 Experimental aerial variogram for TOP_SELE facies.

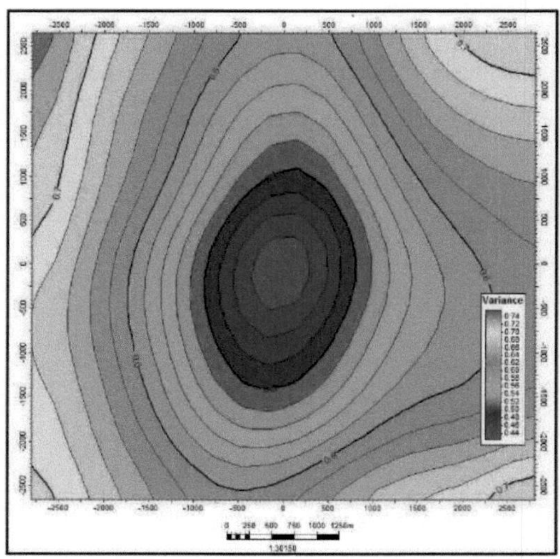

Source: Own elaboration

Fig.4.30 Experimental aerial variogram for facies in Mb4

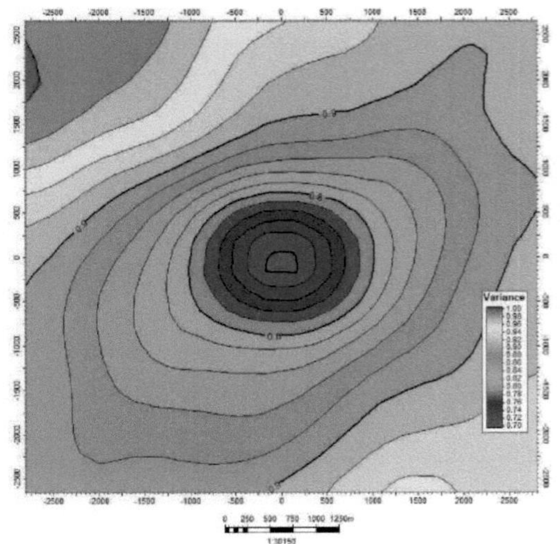

Source: Own elaboration

Fig.4.31 Experimental aerial variogram for facies in TOP_MAUREEN

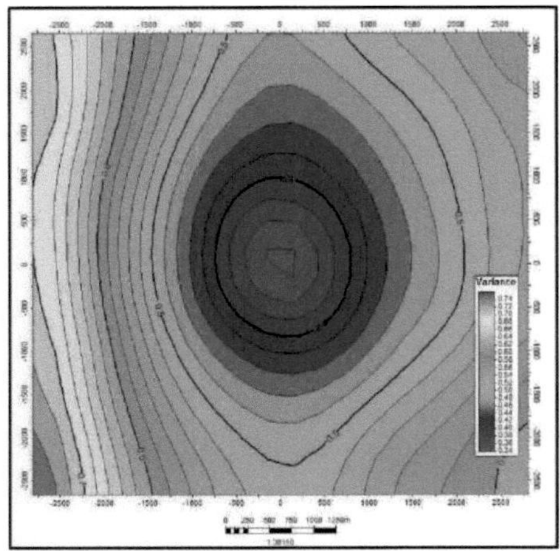

Source: Own elaboration

Fig.4.32 Modeling result for the S3-Thor interval

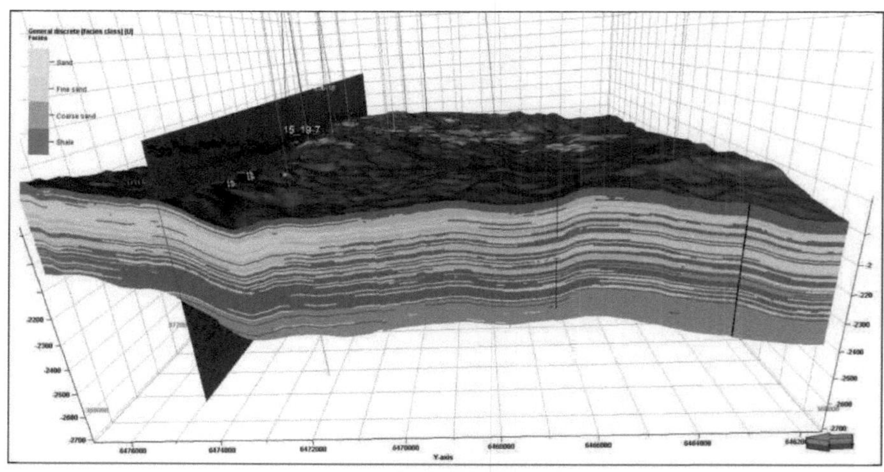

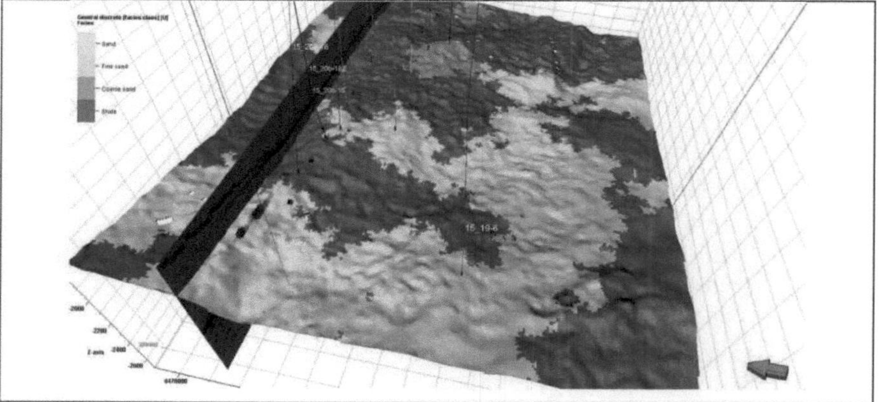

Source: Own elaboration

4.8. Analysis of results and validation of hypotheses

For the validation of hypotheses there are: qualitative tools that depend on the geological knowledge and experience of the modeler; there are also quantitative tools that are based on the distribution of histograms of the calculated properties. For the purpose of this work, both are used because of their complementarity. Clearly it can be observed in Fig.4.33 a good distribution in the proportion in the histogram that has a great approximation in the percentages present of the facies interpreted from the electrical logs, those used in the scale up process and those obtained after the modeling in the whole grid. This symmetry and approximation of values makes it possible to accept the hypothesis as valid.

The obtained result can be contrasted with different calculations, applying modifications in the variables and thus to see the influence of the different parameters in the result of the distribution of the facies, although this procedure is done by removing variables, in a deeper and more specialized study phase it can be done with alternative geological cases or modifying small parameters.

For case A as shown in Fig.4.34, the facies properties are Sand=36%, Fine Sand=18%, Coarse Sand =10%, Shale = 36%, without aerial maps of horizontal variograms, and without RMS variables in cokrigging, but with vertical variogram, in spite of that the distribution in the histogram is relatively acceptable in approximation of values; however, it is possible to appreciate in vertical form abnormal oscillations to the depositional system identified in the area.

For case B, as shown in Fig. 4.35, the same percentage of facies distribution was used as in the previous case, the algorithm used RMS as cokriging variable, however the facies distribution in the histogram shows greater variation, the focused proportion is due to the resolution of the seismic attribute, the abrupt oscillations of facies do not correspond to the stratigraphic transitions identified in the area, but to amplitude anomalies poorly represented or overestimated in importance.

For the calculation of case C, as shown in Fig.4.36, we proceeded to not use the vertical variogram, neither the aerial variogram maps nor any type of variable in cokrigging, respecting the proportions of the facies in case A, the final distribution in the histogram shows a good representation of the values, however in the visual inspection we can appreciate an erratic type without shape or geological representation in the area, this type of facies distribution is the net random type of the algorithm, in which there is no geological criterion.

Fig.4.33 Comparison histogram

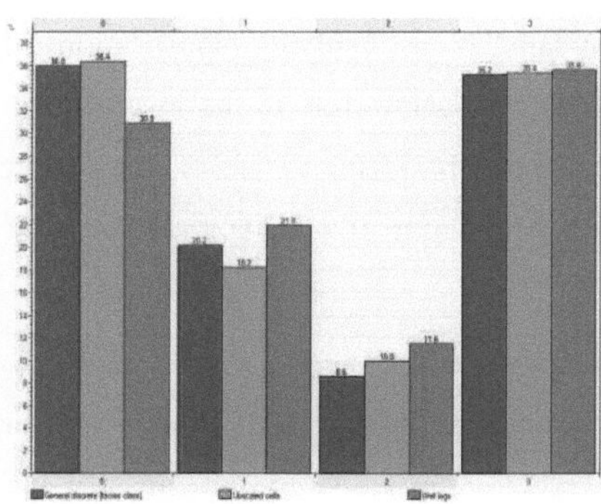

Source: Own elaboration

Fig.4.34 Comparison calculation case A

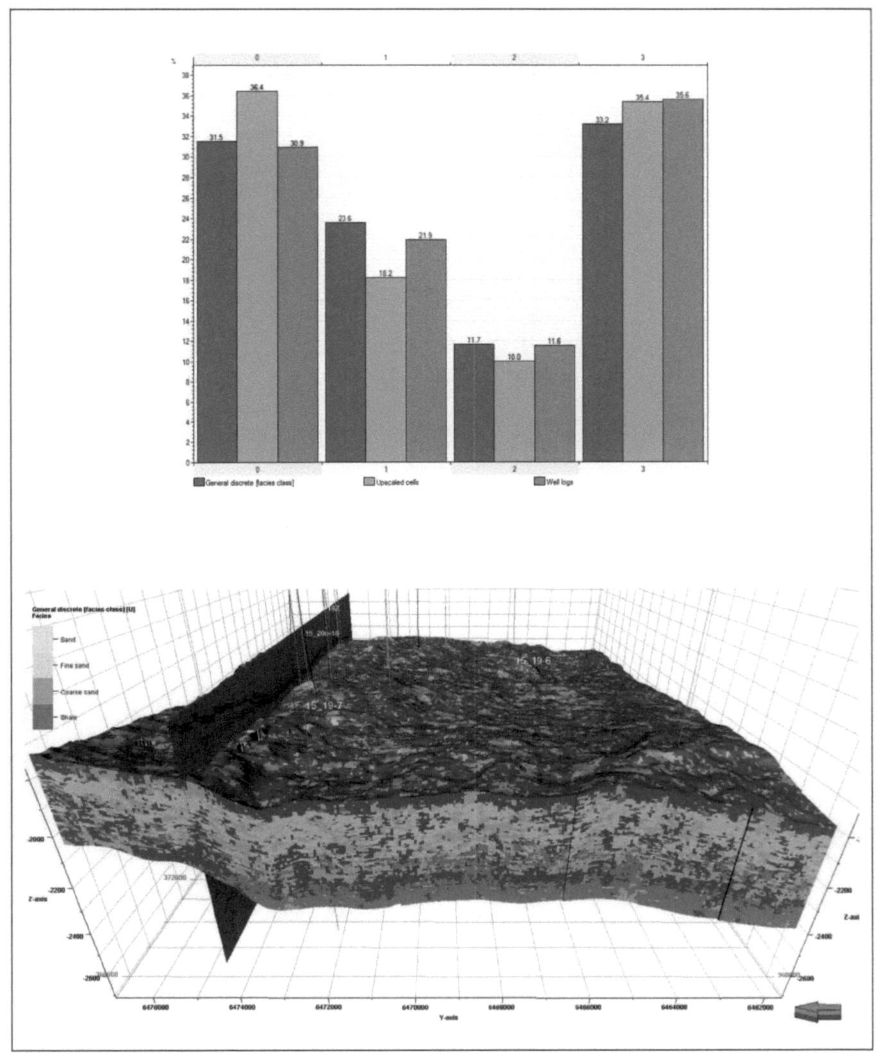

Source: Own elaboration

Fig.4.35 Comparison calculation case B

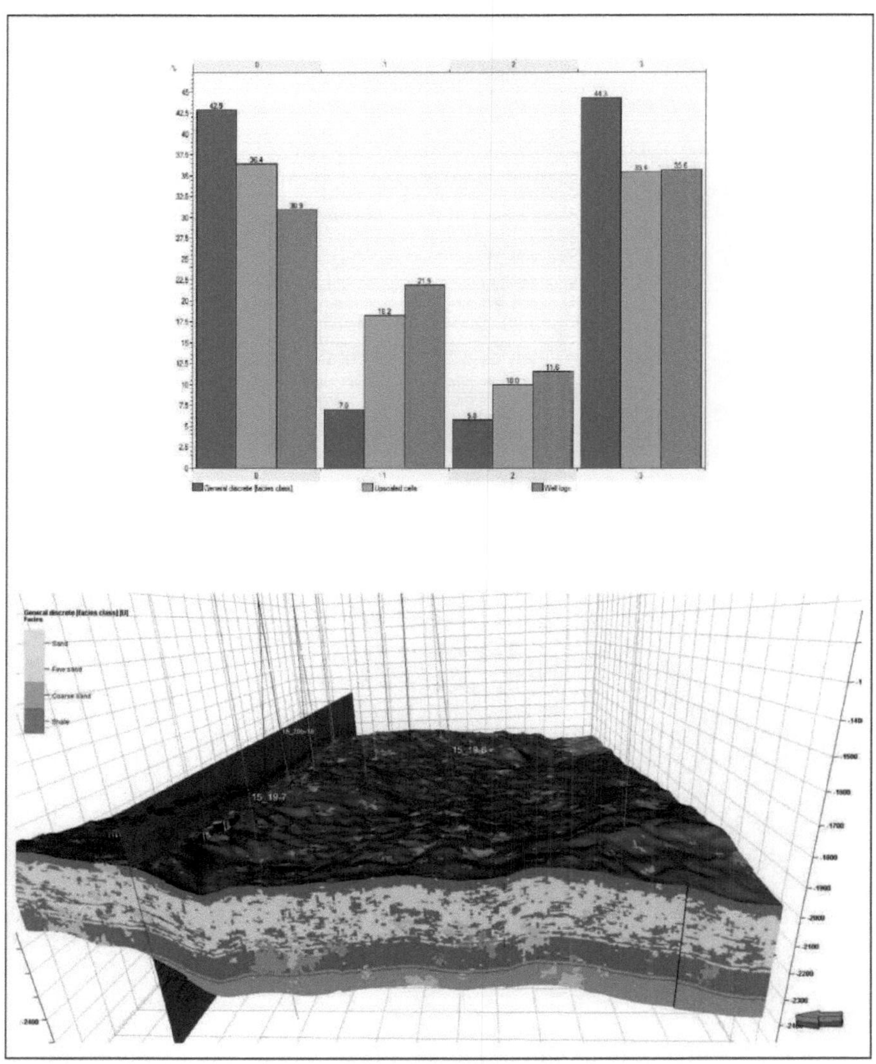

Source: Own elaboration

Fig.4.36 Comparison calculation case C

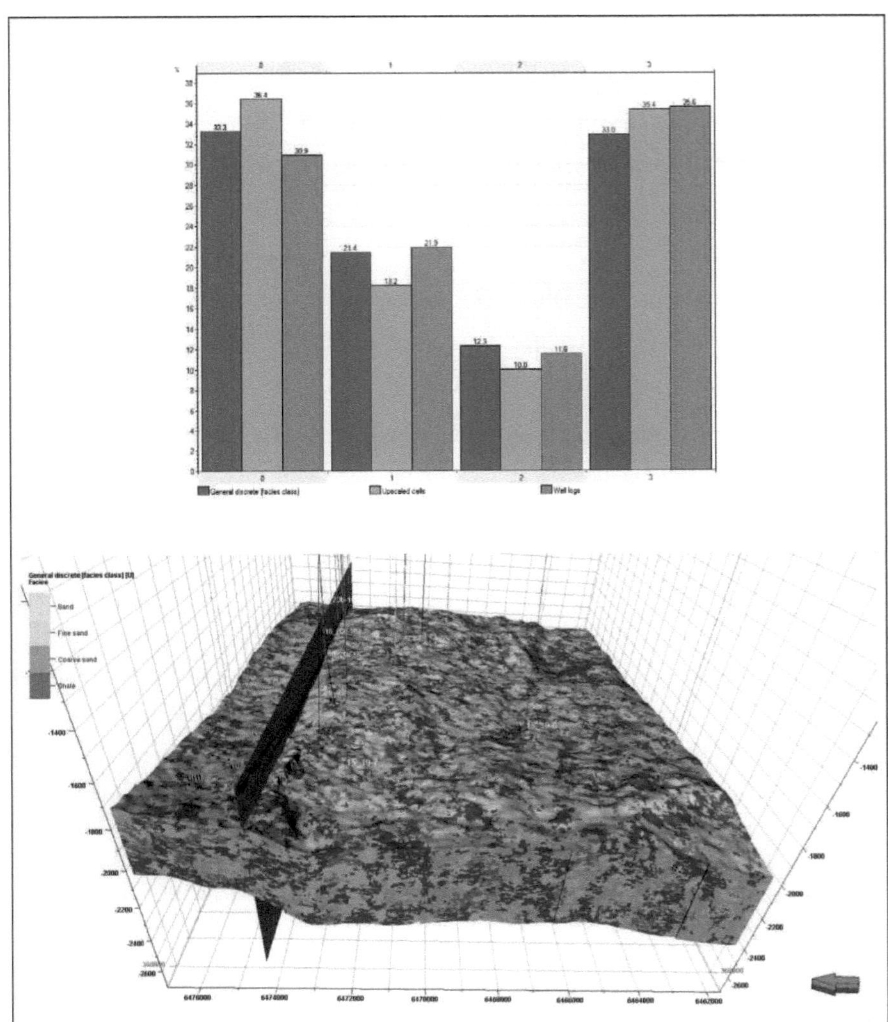

Source: Own elaboration

4.9. Conclusions

Through the present work it can be concluded that:

The optimal vertical resolution of representation is given by the vertical experimental variogram of the wells, taking into account the interpreted facies as a basis for calculation, in a visual inspection it is possible to verify the low distortion and very good representation in the interval of interest.

The best way to attenuate abrupt oscillations is by cokrigging, so that seismic attributes can influence the distribution trend of selected facies susceptible to seismic resolution.

As long as the faults do not generate secondary diagenesis, they can be left aside, since there is no distortion in the facies under evaluation for distribution analysis.

It is extremely important to have a reliable velocity model since it is the link to convert time to depth and vice versa. Although this process is not contemplated in the present work, a small variation in the velocity estimation generates a great uncertainty in the identification of the guide reflector and formation tops in the structural model, causing distortion in the model.

Stratigraphic data are very important in order to integrate geological information obtained in the borehole through drilling reports and electrical logs with their equivalence in the structural model, this ensures the coherent contextualization of the electrical logging information with the seismic expressions and their sedimentary environment.

4.10. Recommendations

The influence of facies interpretation based on electric logs can be improved by using mineralogical inversion and instead of using defined facies intervals, performing the evaluation based on lithology percentages would improve the ability to perform sensitivity analysis for lithofacies prediction.

Frequency attributes could be used to improve the visualization and use a deterministic algorithm taking into account more defined paleoforms based on the visualization technique called RGB blending.

Homogeneity in the terms of the stratigraphic description would improve the identification of reflectors, since different bibliographies use different names for the same stratigraphic elements, this discrepancy generates uncertainty in the correlation of sedimentary bodies.

The histograms only refer to a global comparison, if a more exhaustive validation of the model is required, other more advanced techniques have to be used and in turn an analysis of the minimum number of realizations in which the certainty of the model can be quantified.

Only lithological facies were considered, however, injectites are a very important factor to consider since they are responsible for the migration and flow of fluids, however, being of subseismic resolution, more advanced techniques have to be used.

Since the entire model had to be built and there were no interpreted surface data, it is recommended to take the results obtained as susceptible to seismic reinterpretation and modification of the formation tops used as guide reflectors.

The economic impact of poor facies distribution becomes visible at the time of petrophysical modeling and later in the simulation process, production profile prediction and reserve quantification, so it is recommended to evaluate this influence in these instances, through a much deeper integrated work.

BIBLIOGRAPHY .

Al-Anezi, K., Kumar, S., & Eibad, A. (2013). *Geostatistical Modeling with Seismic Characterization of Wara/Burgan sands-Minagish Field-West Kuwait.* Abu Dhabi UAE: Society of Petroleum Engineers SPE 166046.

Ampilov, Y. (2010). *Seismic Interpretation to Modelling and Assessment of Oil and Gas Fields.* HOUTEN The Netherlands: EAGE Publications.

Catuneanu, O. (2006). *Principles of Sequence Stratigraphy.* Amsterdam, The Netherlands: Elsevier.

Chopra, S., & Marfurt, K. (2007). *Seismic Attributes for Prospect Identification and Reservoir Characterization* (Vol. Geophysical Developments Series No. 11). Tulsa, OK U.S.A: SEG Society of Exploration Geophysicists.

Coburn, T., Yarus, J., & Chambers, L. (2006). Geostatistics and stochastic modeling: Bridging into the 21st century. In T. Coburn, J. Yarus, & L. Chambers, *Stochastic modeling and geostatistics: Principles, methods, and case studies* (Vol. Computer Applications in Geology 5, pp. 3- 9). AAPG American Association of Petroleum Geologists.

Davies, R., Posamentier, H., Wood, L., & Cartwright, J. (Edits.) (2007). *Seismic Geomorphology: Applications to Hydrocarbon Exploration and Production* (Vol. Special Publications 277). London: Geological Society.

Davies, R., Posamentier, H., Wood, L., & Cartwright, J. (2007). *Seismic Geomorphology: Applications to Hydrocarbon Exploration and Production* (Vol. Special Publications 277). London: Geological Society.

Doyen, P. (2007). *Seismic Reservoir Characterization An Earth Modelling Perspective* (Vol. Education Tour Series). Houten The Netherlands: EAGE Publications.

Dubrule, O. (2003). *Geostatistics For Seismic Data Integration In Earth Models.* HOUTEN The Netherlands: EAGE Publications.

Fluckiger, S. D., Hennes, A. M., Zawila, J. S., & Hofmann, M. H. (2015). *Predicting Reservoir Heterogeneity in the Upper Cretaceous Frontier Formation in the Western Power River Basin- An Integrated Stratigraphic, Sedimentologic, Petrophysical and Geophysical Study.* San Antonio Texas USA: Society of Petroleum Engineers SPE 178542 MS.

Gluyas, J. G., & Hichens, H. M. (2003). *United Kingdom Oil and Gas Fields* (Vol. Commemorative Millennium Volume). London, UK: GEOLOGICAL SOCIETY.

Herron, D. A. (2011). *First Steps in Seismic Interpretation*. Tulsa, OK, USA: Society of Exploration Geophysicists.

Jialiang, H., Khalid, O., Johan, W., Mohamed, M., & Neves, F. (2015). *A case study of integrated seismic and well statistics in modeling of deep reservoirs*. Abu Dhabi UAE: Society of Petroleum Engineers SPE-177625.

Le Blevec, T., Dubrule, O., Cedric, J., & Hampson, G. (2016). *Building more realistic 3D facies indicator models*. Bangkok: Paper presented at the International Petroleum Technology Conference.

McKie, T., Rose, P., Hartley, A., Jones, D., & Armstrong, T. (2015). *Tertiary DeepMarine Reservoirs of the North Sea Region* (Vols. Special Publications, 403). London: Geological Society.

Myers, D. E. (2006). Reflections on geostatistics and stochastic modeling. In T. Coburn, J. Yarus, & L. Chambers, *Stochastic modeling and geostatistics: Principles, methods, and case studies* (Vol. Computer Applications in Geology 5, pp. 11-22). AAPG American Association of Petroleum Geologists.

Pelgrain, A., Cosetino, L., Cabrera, J., Jimenez, T., & Bellorin, O. (2002). *Geologically oriented geostatistics: an integrated tool for reservoir studies*. Mexico: Paper presented at SPE international Petroleum Conference.

Posamentier, H., & Allen, G. (1999). *Siliciclastic Sequence Stratigraphy-Concepts and Applications*. Tulsa, Oklahoma, U.S.A: SEPM Society for Sedimentary Geology.

Pyrcz, J., Gringarten, E., Frykman, P., & Deutsch, C. V. (2006). Representative input parameters for geostatistical simulation. In T. Coburn, J. Yarus, & L. Chambers, *Stochastic modeling and geostatistics: Principles, methods and case studies* (Vol. Computer Applications in Geology 5 volume II, pp. 123137). AAPG American Association of Petroleum Geologists.

Veeken, P. C. (2007). *Seismic Stratigraphy, Basin Analysis and Reservoir Characterization* (Vol. Seismic Exploration Volume 37). The Netherlands: Elsevier.

Veeken, P., & Van Moerkerken, B. (2013). *Seismic Stratigraphy and Depositional Models.*

HOUTEN The Netherlands: EAGE Publications.

Weimer, P., & Slatt, R. M. (2004). *Petroleum Systems of Deepwater Settings* (Vol. Distinguished Instructor Series No7). Tulsa, OK. USA: Society of Exploration Geophysicists.

Zakrevsky, K. E. (2011). *Geological 3d Modelling.* HOUTEN The Netherlands: EAGE Publications.

Zhou, Y., Muggeridge, A., Berg, C., & King, P. (2015). *Quantifying cross flow and its impact on tertiary polymer flooding in heterogeneous reservoirs.* European Symposium on Improved oil Recovery IOR.

GLOSSARY

Deterministic algorithms: Based on defined input data where the final result will be calculated based on a predefined formula and where the final result is unique and depends only on the data.

Anisotropy: A way of measuring whether the variance within a data collection is determined by a direction (measured in azimuth and percent eccentricity).

Gas chimney: Region of escaping gas migrating upward from a hydrocarbon accumulation, appearing as low-amplitude anomalies, chaotic zones recognizable in seismic data.

Classification: organization of data into groups that represent a specific property.

Coherence: measure of the similarity of a seismic waveform through Inline and Xline correlations.

Continuous: A number defined in a rock property with unlimited number of possibilities e.g. porosity, permeability, etc.

Correlation: A way of measuring whether two separate collections of data are related (measured as a percentage).

Standard deviation: The square of the variance and is used to describe how a data distribution varies from the mean.

Dewatering: Process of expelling water out of shales and muds.

Discrete: Code assigned to a property with a limited number of possible options, e.g., facies, bodies, lithologies.

Normal distribution: Means that most of the measurements in a collection of data are close to the mean value while relatively few samples tend toward the extremes on either side.

Seismic facies: The characteristics of a group of reflectors involving amplitude, abundance, continuity, and configuration. A characteristic of seismic waves in their path and their relation to a specific location in a spatial seismic image.

Geostatistics: A branch of applied statistics that emphasizes the geological context of the data and the spatial relationship between the data.

Reservoir heterogeneity: Spatial variability of rock properties in a reservoir that can be produced by geologic processes such as sedimentation, diagenesis, erosion, faulting, etc. A multivariate analysis of geostatistical techniques are commonly used to estimate the level of heterogeneity using borehole and seismic data.

Histogram: A graphical representation of the frequency distribution of the chosen variable(s).

Kriging: A linear system of equations where the variogram values are known parameters and where the weights are unknown parameters.

Facies modeling: Interpolation, calculation or simulation of discrete data such as lithology, bodies or facies.

Stochastic modeling: It is based on a set of input data and where the outcome is going to be calculated in a probabilistic framework. The results will be multiple realizations depending on the input data and a random assignment path.

Geological model: A group of cells with defined volume on the same fault structure and boundaries. Each project could contain many models and each model many 3D grids.

Scale up: The process of assigning values from electrical logs or their attributes to a grid, ready to be used in the facies and petrophysical modeling process.

Transformation: Preparation of a data set into internal data, which meet statistical requirements according to the chosen algorithm.

Trend: Continuous or permanent change in the average value of a property in a 1D, 2D and 3D model.

Variance: A measure of how dispersed a distribution is, calculated as the mean square of each of the numbers in its mean.

Variogram: A quantitative description of the variation of a property as a function of distance of separation between points, plotted graphically as a variance (Y) vs class distance (X) plot.

Buy your books fast and straightforward online - at one of world's fastest growing online book stores! Environmentally sound due to Print-on-Demand technologies.

Buy your books online at
www.morebooks.shop

Kaufen Sie Ihre Bücher schnell und unkompliziert online – auf einer der am schnellsten wachsenden Buchhandelsplattformen weltweit! Dank Print-On-Demand umwelt- und ressourcenschonend produziert.

Bücher schneller online kaufen
www.morebooks.shop

info@omniscriptum.com
www.omniscriptum.com

Printed by Books on Demand GmbH, Norderstedt / Germany